I0887637

"Knowing is not enough; we must apply.
Willing is not enough; we must do."
—Goethe

INSTITUTE OF MEDICINE
OF THE NATIONAL ACADEMIES

Shaping the Future for Health

THE NATIONAL ACADEMIES

Advisers to the Nation on Science, Engineering, and Medicine

The **National Academy of Sciences** is a private, nonprofit, self-perpetuating society of distinguished scholars engaged in scientific and engineering research, dedicated to the furtherance of science and technology and to their use for the general welfare. Upon the authority of the charter granted to it by the Congress in 1863, the Academy has a mandate that requires it to advise the federal government on scientific and technical matters. Dr. Bruce M. Alberts is president of the National Academy of Sciences.

The **National Academy of Engineering** was established in 1964, under the charter of the National Academy of Sciences, as a parallel organization of outstanding engineers. It is autonomous in its administration and in the selection of its members, sharing with the National Academy of Sciences the responsibility for advising the federal government. The National Academy of Engineering also sponsors engineering programs aimed at meeting national needs, encourages education and research, and recognizes the superior achievements of engineers. Dr. Wm. A. Wulf is president of the National Academy of Engineering.

The **Institute of Medicine** was established in 1970 by the National Academy of Sciences to secure the services of eminent members of appropriate professions in the examination of policy matters pertaining to the health of the public. The Institute acts under the responsibility given to the National Academy of Sciences by its congressional charter to be an adviser to the federal government and, upon its own initiative, to identify issues of medical care, research, and education. Dr. Harvey V. Fineberg is president of the Institute of Medicine.

The **National Research Council** was organized by the National Academy of Sciences in 1916 to associate the broad community of science and technology with the Academy's purposes of furthering knowledge and advising the federal government. Functioning in accordance with general policies determined by the Academy, the Council has become the principal operating agency of both the National Academy of Sciences and the National Academy of Engineering in providing services to the government, the public, and the scientific and engineering communities. The Council is administered jointly by both Academies and the Institute of Medicine. Dr. Bruce M. Alberts and Dr. Wm. A. Wulf are chair and vice chair, respectively, of the National Research Council.

www.national-academies.org

COMMITTEE ON THE LONGITUDINAL STUDY OF ASTRONAUT HEALTH

DAVID E. LONGNECKER (*Chair*), Senior Vice President and Corporate Chief Medical Officer and Robert D. Dripps Professor of Anesthesia, University of Pennsylvania Health System, Philadelphia, Pennsylvania

ALFRED F. CONNORS, JR., Charles H. Rammelkamp Professor of Medicine, Case Western Reserve University, and Chair, Department of Medicine, MetroHealth Medical Center, Cleveland, Ohio

ROY L. DEHART, Director, Vanderbilt Center for Occupational and Environmental Medicine, Nashville, Tennessee

R. J. MICHAEL FRY, Retired Head of Cancer Section, Oak Ridge National Laboratory, Indianapolis, Indiana

DANIEL R. MASYS, Director of Biomedical Informatics and Professor of Medicine, University of California, San Diego

VAN C. MOW, Stanley Dicker Professor of Biomedical Engineering and Orthopaedic Bioengineering, and Chair, Department of Biomedical Engineering, Columbia University, New York, New York

TOM S. NEUMAN, Professor of Clinical Medicine and Associate Director, Emergency Medical Services, San Diego Medical Center, University of California, San Diego

THOMAS F. OLTMANNS, Edgar James Swift Professor of Arts and Sciences, Department of Psychology, Washington University, St. Louis, Missouri

RUSSELL B. RAYMAN, Executive Director, Aerospace Medical Association, Arlington, Virginia

WALTER ROBINSON, Associate Professor of Pediatrics and Medical Ethics, Division of Medical Ethics, Harvard Medical School, Boston, Massachusetts

ELAINE RON, Senior Investigator, Radiation Epidemiology Branch, National Cancer Institute, National Institutes of Health, Rockville, Maryland

CAROL SCOTT-CONNER, Chair, Department of Surgery, University of Iowa Hospitals and Clinics, Iowa City, Iowa

M. RHEA SEDDON, Assistant Chief Medical Officer, Vanderbilt Medical Group, Vanderbilt University Medical Center, Nashville, Tennessee

DEBORAH ZUCKER, Assistant Professor of Medicine, Tufts University and Clinical Investigator, Division of Clinical Care Research, New England Medical Center, Boston, Massachusetts

Study Staff

FREDERICK J. MANNING, Study Director

NATASHA S. DICKSON, Senior Project Assistant

BENJAMIN HAMLIN, Research Assistant

Institute of Medicine Staff
ANDREW POPE, Director, Board on Health Sciences Policy
MELVIN H. WORTH, JR., Scholar-in-Residence
TROY PRINCE, Administrative Assistant, Board on Health Sciences Policy
CARLOS GABRIEL, Financial Associate

INDEPENDENT REPORT REVIEWERS

This report has been reviewed in draft form by individuals chosen for their diverse perspectives and technical expertise, in accordance with procedures approved by the National Research Council's Report Review Committee. The purpose of this independent review is to provide candid and critical comments that will assist the institution in making its published report as sound as possible and to ensure that the report meets institutional standards for objectivity, evidence, and responsiveness to the study charge. The contents of the review comments and draft manuscript remain confidential to protect the integrity of the deliberative process. We wish to thank the following individuals for their participation in the review of this report:

John R. Ball, American Society for Clinical Pathology
John Boice, International Epidemiology Institute
F. Andrew Gaffney, Vanderbilt University
Thomas A. Louis, Johns Hopkins University
Jay H. Lubin, National Institutes of Health
Jonathan D. Moreno, University of Virginia
Deborah J. Wear-Finkle, Maine Cognitive Therapy Center

Although the reviewers listed above have provided many constructive comments and suggestions, they were not asked to endorse the conclusions or recommendations nor did they see the final draft of the report before its release. The review of this report was overseen by **ROBERT M. EPSTEIN,** Harold Carron Professor of Anesthesiology Emeritus at the University of Virginia, appointed by the Institute of Medicine, who was responsible for making certain that an independent examination of this report was carried out in accordance with institutional procedures and that all review comments were carefully considered. Responsibility for the final content of this report rests entirely with the authoring committee and the institution.

Preface

As part of its ongoing commitment to the nation's space program, NASA's medical leadership asked the Institute of Medicine (IOM) to review specific aspects of the scientific basis, policies, and procedures associated with the Longitudinal Study of Astronaut Health (LSAH). NASA created the LSAH in 1992 to address a variety of issues, including both the health of astronauts during space flight and the longer-term health issues that might be associated with space flight and flight training.

The IOM Committee on the LSAH held most of its deliberations at the new IOM facilities in Washington, DC, where the group pondered a variety of health care issues related to space flight, astronaut training, and subsequent astronaut health. We spent many hours developing an in-depth understanding of the LSAH, the major risk factors related to space flight and flight training, and the subsequent health of astronauts. The highlight of the committee's experiences took place at the Kennedy Space Center (KSC) on January 15-16, 2003, when the committee met with numerous NASA scientists associated with the LSAH, all of whom were gathered at KSC for the scheduled launch of STS-107, the *Columbia* orbiter flight devoted to life sciences. The committee heard numerous scientific presentations on January 15, including those by the flight surgeons associated with the STS-107 *Columbia* crew. After an informative session of scientific presentations and deliberations, the committee was escorted to a night viewing of the launch site, and early Thursday morning, January 16, we attended the NASA prelaunch briefing and the subsequent launch of STS-107. Although there were occasional intervals of concern during the last 24 hours of the launch count-down, in general the launch cycle was almost routine; some described it as one of the smoothest launch cycles in recent years. In mid-morning, STS-107 lifted off (perhaps "leapt off" would be more accurate) the launch pad and disappeared into a gorgeous blue sky within five minutes. The flight controllers, crew, NASA administrators and staff, the throngs of visitors, and the committee were thrilled by this sight. The realities of the committee's assignment were brought into sobering focus on the morning of February 1, when the image of

Columbia returning to the earth's atmosphere suddenly became multiple images over the clear Texas skies. For me, and for many others on the committee, both the launch and the disintegration of *Columbia* are forever printed into our visual memories.

The events of February 1 served to remind the committee of the perilous nature of space flight, and brought back memories of *Challenger* in 1986 and *Apollo 1* in 1967. In all, the issues we address in this report are important, vital, and meaningful. However, beyond the long-term issues of thyroid function, behavioral medicine, cataracts, and cancer, all of which are addressed in this review, there remains the harsh reality that space flight is an inherently risky endeavor and space flyers are at risk both during training and in flight. Our recommendations address ways to mitigate at least some of these risks where possible or to compensate for health risks that cannot be anticipated or eliminated. Our committee dedicates this volume, and our many long hours of meetings, reading, analysis, deliberation, and writing, to Rick Husband (Commander of STS-107), William McCool (Pilot), Kalpana Chawala (Flight Engineer), David Brown, M.D. (Mission Specialist), Laurel Clark, M.D. (Mission Specialist), Michael Anderson (Payload Commander), and Ilan Ramon (Payload Specialist). *Requiescant in pace.*

David E. Longnecker, *Chair*
Committee on the Longitudinal Study of Astronaut Health

Contents

List of Tables

Executive Summary

The career of an explorer is risky, and it is chosen by individuals who acknowledge and accept risks beyond those of ordinary daily living. As the disintegration of the space shuttle *Columbia* upon reentry into the earth's atmosphere in February 2003 so vividly demonstrated, space travel has unique risks. In addition to the tremendous engineering challenges entailed in getting space travelers launched and returned safely, biomedical information collected by the National Aeronautics and Space Administration (NASA) and the Soviet and Russian space programs has revealed that living in space can produce profound physiological and clinical changes. Much less is known about potential longterm effects of space flight or the overall occupational risks of being an astronaut. NASA physicians began thinking about a longitudinal study as early as the late 1970s, and the Longitudinal Study of Astronaut Health (LSAH) was approved in 1992. Ten years later, NASA's Chief Health and Medical Officer asked the Institute of Medicine (IOM) for help in assessing the study and making any necessary midcourse corrections.

ROLE OF THE INSTITUTE OF MEDICINE

A prior IOM report entitled *Safe Passage* (IOM, 2001c) is recommended as background reading for this study. Despite the fact that it focused on the immediate dangers to the health and safety of astronauts aboard a future mission to Mars, it examined many issues of relevance to the present study, including the role of the astronauts as research subjects and the need for a comprehensive health care system for astronauts.

Presently the IOM, through activities including studies and workshops undertaken at the National Academies under the auspices of its standing Committee on Aerospace Medicine and the Medicine of Extreme Environments (CAMMEE), provides NASA's Chief Health and Medical Officer independent technical advice relevant to aerospace medicine, including medical care of space travelers. In October 2002 NASA's Chief Health and Medical Officer wrote a letter to the IOM project officer that described some tentative findings from a recent analysis of the LSAH database by scientists at the Johnson Space Center (JSC) and requested that CAMMEE examine the LASH and make appropriate medical, scientific, and administrative recommendations for improving the study, as well as recommendations relative to the data trends identified to date. CAMMEE in turn organized the present Committee on the Longitudinal Study of Astronaut Health (CLSAH), which convened for the first time in conjunction with the January 2003 meeting of CAMMEE. NASA had performed some further analysis of the LSAH database in the interim, and after presentation of those analyses, CLSAH's task was revised and expanded to yield the following charge to the committee:

> Examine NASA's Longitudinal Study of Astronaut Health (LSAH) and make appropriate medical, scientific, and administrative recommendations for improving the study, as well as recommendations relative to the data trends identified to date, inclusion of astronauts from NASA's international partners, appropriate follow up of findings, and medical care of current and former astronauts, mission specialists, and other space travelers. In so doing the committee will address the potential relevance of lessons learned from historical exposures such as agent orange, radiation among veterans, and industrial beryllium to the configuration of the LSAH with regard to its usefulness in identifying health risks.

GOALS AND DESIGN OF THE LSAH

According to the most recent published description of the LSAH (Hamm et al., 2000), the primary aim of the LSAH is "to investigate and describe the incidence of acute and chronic morbidity and mortality of astronauts and to determine whether the unique occupational exposures encountered by astronauts are associated with increased risks of morbidity or mortality."

The primary focus of the study is the 312 men and women who have been selected as NASA astronauts since the space program began in 1959. All active astronauts participate in the study. Astronauts who have retired or otherwise left NASA are invited to continue in the study, and their participation rate is high (varying from 61 percent to 88 percent over the nine years between 1993 and 2001).

The study also collects health and medical data from a non-astronaut comparison group of JSC employees matched for sex, age, and body mass index (BMI). The study design calls for a 3:1 ratio of comparison participants to astronaut participants, and in January 2003 the comparison group totaled 928.

The primary data for the LSAH are obtained from medical records maintained at the JSC clinics. Annual health evaluations are required of active astronauts and are offered to inactive astronauts. These evaluations consist of a medical history, physical examination, laboratory tests, medical images, and other diagnostic tests. Non-astronaut employees who are participating in this study are offered evaluations every other year. Reports and documentation of interim medical care are obtained as part of the evaluation. These evaluations are referred to as "physical exams" throughout the report, though they clearly include much more. Other study data are obtained from interim visits to the JSC clinics for sick calls, and, for the astronauts, from pre- and post-flight physical exams, medical debriefings following flights, inflight experimental data, and reports of inflight medical events. A questionnaire designed to capture lifestyle factors and health risk data is now mailed to all new participants when they enter the study and every two years thereafter. Biannual searches are done for death certificates of all participants who miss a scheduled physical exam and cannot be contacted by mail or telephone. Copies of autopsy reports and hospital death summaries to support death certificate data are obtained whenever they are available.

FINDINGS TO DATE

Several analyses of the LSAH database have been published in peer-reviewed journals, the earliest a 1993 report on astronaut mortality from 1959 though 1991 that also addressed the hypothesis that astronauts are at increased risk for fatal cancers. Updated and expanded analyses were published in 1998 and 2000; a paper devoted to cataracts in astronauts was published in 2001; and in meetings held early in 2003 the committee was briefed by JSC scientists on more recent analyses of morbidity and mortality. Chapter 2 summarizes the three published studies as well as the briefings, but since the latter built on and were consistent with the earlier reports, only the latest analyses are reported in this summary.

Overall mortality has been significantly higher for the astronaut group in every analysis. Data presented to the committee in January 2003, just prior to the loss of the space shuttle Columbia and its crew of 7, showed 29 deaths among the 312 astronauts in the LSAH database and only 17 deaths among the 912 comparison participants. Accidental deaths, including 8 in spacecraft losses, accounted for 20 of the astronaut deaths (versus only 2 in the comparison group). The groups did not differ significantly in mortality from any other cause.

LSAH data on cataract incidence was combined with individual radiation exposure data from 295 astronauts in a study by Cucinotta and colleagues (Cucinotta et al., 2001; Cucinotta, 2003) which suggested increased incidence and earlier appearance of cataracts in astronauts exposed to higher amounts of space radiation (>8milliSieverts). A follow up study is using digital photography and computer image analysis to better quantify cataract incidence and progression using a group of current and former military pilots as controls.

Because of the known association of some cancers with radiation exposure, surveillance of astronauts for malignancies was planned from the beginning of the LSAH. Craig Fischer briefed the committee on the comparison of cancer incidence among the astronauts (Fischer, 2003), the LSAH comparison participants, and an age- and sex-matched sample of the National Cancer Institute's Surveillance, Epidemiology and End Results (SEER) database. Fourteen cases of cancer (excluding 33 cases of non-melanoma skin cancer) were diagnosed among the 312 astronauts followed from 1959 to the present. This is 59 percent higher than the comparison group per person/year (not statistically significant), but 46 percent lower per person/year than the SEER data (statistically significant).

The LSAH database also played a role in correcting a serious problem involving excessive iodinization of space shuttle drinking water. A physician monitoring the health and safety of four astronauts in a ground-based test of space shuttle life-support systems discovered marked elevations in thyroid stimulating hormone (TSH)—an indicator of potentially abnormal thyroid gland function—in all four after only 30 days of the 90-day test. She noted that iodine introduced into the test subjects' drinking water as a bacteriocide was increasing the astronauts' iodine intake to levels long recognized as detrimental to thyroid function. Installation of anion exchange filters sharply reduced the iodine concentration at the ground study tap, and the astronauts' TSH levels gradually returned to normal. A retrospective review of LSAH data showed no significant difference between the astronauts and the comparison participants in clinical thyroid disease but that elevations of TSH during flight had been common, with gradual return to normal after return to earth. Anion exchange filters are now a standard component of the drinking water systems on the space shuttle, and transient elevations in TSH no longer occur.

PROBLEMS IN DESIGN AND EXECUTION

To obtain an unbiased estimate of risk, the astronauts and their comparison group controls should (1) be equivalent at baseline in all factors that influence risk of disease or adverse health outcomes; (2) have equivalent exposures in day-to-day life except for those related to spaceflight or preparation for space flight; (3) have equivalent monitoring for disease by observers blinded to whether or not they were exposed to spaceflight or spaceflight preparation; and (4) participate fully from study entry to the outcome of interest. Like many expensive, long-running epidemiological studies, the LSAH has had to make a number of compromises. Chief among these have been a less than ideal match between comparisons and astronauts on a number of other potentially relevant physical and psychosocial variables, increasing disparities in the surveillance of health problems in the astronaut and comparison groups, and a lower participation/followup rate of the comparison group. The proposed inclusion of astronauts and cosmonauts from NASA's international partners into the LSAH would only add to these problems for interpretation, although some of the data from longterm missions would be valuable independent of its contribution to the LSAH.

RECOMMENDATIONS FOR CHANGE

Implementing the following recommendations, which subsume many offered by the LSAH staff, will inevitably involve additional expenditures, but the committee believes they are essential for the validity of the data gathered through the LSAH and ultimately for the creation of a safer space travel program.

Recommendation 1

NASA should recognize that the LSAH can and should serve the two separate and potentially conflicting goals of occupational surveillance of the health of current and former astronauts and research into the long-term health risks associated with manned space flight (and to make these activities safer for future astronauts).

a) For the surveillance portion of the survey, participation of the astronauts is mandatory; for the research portion it is voluntary. Consequently, for the research portion, the astronauts need to sign an up-to-date informed consent document, and the research portion of the study should be reviewed on a regular basis by an IRB.

b) The database should be reviewed no less often than annually by LSAH staff, and analyses should be conducted for areas of potential risk, e.g., cancers, hearing loss, cataracts, bone strength. The committee is not convinced, given the low power of the study, that traditional "statistical significance" should be the sole trigger for concern, so in addition, it recommends that routine surveillance for unexpected and sentinel events be carried out by the oversight committee described below.

c) There should be a formal mechanism for flight surgeons to discuss both among themselves, and with those involved in the LSAH, any outlier or sentinel events, so that clinical suspicions are shared and checked for generality; such a system should complement the database surveillance system described above.

d) More information should be provided to participants on emerging findings and possible risks (possibly via their examining physician). The current newsletter system could be supplemented by a clinical synopsis with an expert commentary as key findings are published.

e) A formal process should be established to determine and implement corrective actions that follow from database surveillance or adverse event reporting. This process should enable the most learning to occur so that current and future astronauts are enabled to lead less risky lives, at least in their calling as explorers.

f) The Health and Lifestyle Questionnaire should be regularly reviewed with outside experts and updated as recommended.

Recommendation 2

NASA should recognize that no comparison group can meet every goal of the LSAH. Although use of the existing comparison group can be improved (see below), other hypothesis-specific comparison groups will be needed for definitive assessment of specific risks identified in the astronaut population. The comparison group should be seen primarily as a means to detect possible anomalies. Only after anomalies are identified can the most appropriate control group be identified and a definitive assessment of risk made. Specific suggestions for the current comparison group are:

a) The ratio of three comparison participants for each astronaut selected should be maintained. JSC contractor (e.g., Wyle Labs) personnel should be added to the comparison participant pool if the civil servant population can no longer provide adequate matches for new astronaut classes.

b) NASA should continue to seek international partner astronauts' medical data, but we do not recommend pooling such data with the LSAH data. Greater priority should be given to more thorough data gathering from the existing participant groups.

Recommendation 3

NASA should take steps to increase the quantity and improve the quality of the data collection and management of the data of the LSAH. The Committee was concerned by the marked variation in the content of the screening examinations that the existing LSAH groups (astronauts, retired astronauts, civil servants, and retired civil servants) are currently receiving, by the extent of missing data in some areas, and by the lack of justification for including some screening examinations and omitting others. These issues should be reviewed in accordance with the following principle: Exact or near-exact similarity of examination content in all four groups is more important than close similarity of examination frequency. Specific steps might include:

Data Collection

a) Pay travel expenses for comparison participants who no longer work at JSC and live outside the Houston area. Former astronauts who live outside the Houston area are already reimbursed for travel expenses, as are active astronauts and JSC civil servant participants, if they incur any expenses.

b) Offer to pay for an equivalent examination to be performed at a site convenient to comparison participants outside the Houston area. Occupational health clinics at other NASA centers, Federal Aviation Agency medical examiners, or private primary care providers could be given a standard protocol.

c) Institute a publicity campaign to notify LSAH participants of the new benefit of receiving physical exams and laboratory tests comparable to those of the astronauts.

d) Implement a more active program to identify and contact individuals who miss a scheduled physical and ascertain reasons for non-participation.

e) Implement a more active program to obtain medical records from private health care providers. The JSC Occupational Health Clinic provides no treatment for former employees. Participants are simply told the results of their physical exams and lab tests and referred to their private physicians for treatment of any suspected conditions. Participants are asked to forward the records of those subsequent appointments with their private providers.

f) Inflight radiation dosimetry should be state of the art and carefully recorded in the LSAH database, along with exposures of both astronauts and comparison participants in diagnostic and therapeutic settings on earth; Analyses should be carried out by categories of "radiation dose" wherever possible.

The addition of the following would enhance the value of the study:

g) Mental health data should be added to the LSAH database.
h) Biological specimens should be stored for future tests and studies.

Data Management

The Committee recommends several changes in the oversight structure for the LSAH:

i) A standing oversight committee should be established with the participation of ex-astronauts, the public, scientists of various disciplines, and independent external reviewers. The expertise needed by such a committee includes biostatistics, clinical medicine, etc. Principal activities of such an oversight committee should be review of the methods used to acquire and analyze the data, surveillance of the data set for unexpected events, and evaluation of plans for reacting to these events. In addition, this oversight committee should set up procedures for site review of the performance of the study analogous to that performed by clinical research organizations.

j) At least one ex-astronaut and one or more non-NASA biomedical scientists should be added to the existing LSAH Executive Committee.

k) Additional professionals (e.g., epidemiologist) and staff should be hired as necessary to keep the database current and meet the new review and reporting requirements described above.

Finally, the committee addressed the need for NASA to have a policy addressing the practical consequences of discovering that a career as an astronaut, or the experience of space travel, leaves astronauts at increased risk for an adverse health effect. Of particular concern is the case where the effect does not become obvious during or immediately after a space flight, but instead develops sometime after the astronaut leaves active duty and is no longer provided medical care by NASA. After reviewing the history and policies of the Departments of Energy, Defense, and Veterans Affairs in somewhat analogous cases involving beryllium, nuclear weapons tests, and Agent Orange, respectively, the committee's final recommendation was to reiterate a suggestion of the committee that authored *Safe Passage* (IOM, 2001c).

Recommendation 4

NASA should assume responsibility for the lifelong health care of its active and former astronauts.

1

Introduction

The career of an explorer is risky, and it is chosen by individuals who acknowledge and accept risks beyond those of ordinary daily living. As the disintegration of the space shuttle *Columbia* upon reentry into the earth's atmosphere in February 2003 so vividly demonstrated, space travel has unique risks. In addition to the tremendous engineering challenges entailed in getting space travelers launched and returned safely, biomedical information collected by the National Aeronautics and Space Administration (NASA) and the Soviet and Russian space programs has revealed that living in space can produce profound physiological and clinical changes. These changes include the loss of calcium and other minerals from bone, decrease in skeletal muscle mass, decreased or altered absorption of nutrients in the gastrointestinal tract, disturbed fine motor control, increased risks of renal calculi, anemia, and depressed immune system function (Nicogossian et al, 1993; IOM, 2001c). It is now clear that humans can survive and perform acceptably in space for periods of as long as a year despite these changes and that most but not all of the changes are reversible upon return to earth.

Much less is known about the potential longterm effects of space flight that are not apparent in the inflight and immediate postflight medical data collected to date, nor is much known about the overall risks of being an astronaut. NASA physicians began contemplating a longitudinal study as early as the late 1970s. In 1980, they convened a panel of eminent epidemiologists to help design a protocol for retrospectively examining basic physiological data from the relatively small number of astronauts who had flown in space by then and comparing those data with similar data from a group of ground-based employees selected retrospectively to match the living astronauts. However, the current prospective study, the Longitudinal Study of Astronaut Health (LSAH), was not approved by the Human Research Policy and Procedures Committee of the Johnson Space Center (JSC) until 1992. Ten years later, NASA's Chief Health and Medical Officer asked the Institute of Medicine (IOM) for help in assessing the study and making any necessary midcourse corrections.

ROLE OF THE INSTITUTE OF MEDICINE

A prior IOM report entitled *Safe Passage* (IOM, 2001c) is recommended as background reading for this study. Despite the fact that it focused on the immediate dangers to the health and safety of astronauts aboard a future mission to Mars, it examined many issues of relevance to the present study, including the role of the astronauts as research subjects and the need for a comprehensive health care system for astronauts.

Presently, the IOM, through activities including studies and workshops undertaken at the National Academies under the auspices of its standing Committee on Aerospace Medicine and Medicine in Extreme Environments (CAMMEE), provides NASA's Chief Health and Medical Officer independent technical advice relevant to aerospace medicine, including medical care of space travelers. A May 2001 CAMMEE meeting included a presentation by scientific staff from the JSC on the LSAH that stimulated considerable discussion and a request by the CAMMEE for additional information at a future meeting. In early fall of the same year, NASA's Chief Health and Medical Officer wrote a letter to the IOM project officer that described some tentative findings from a recent analysis of the LSAH database by JSC scientists and requested that CAMMEE examine the LSAH and make appropriate medical, scientific, and administrative recommendations for improving the study, as well as recommendations relative to the data trends identified to date. CAMMEE in turn organized the present Committee on the Longitudinal Study of Astronaut Health (CLSAH), which convened for the first time in conjunction with the January 2003 meeting of CAMMEE. NASA had performed some further analysis of the LSAH database in the interim, and after presentation of those analyses, CLSAH's task was revised and expanded to yield the following charge to the committee:

> An ad hoc subcommittee formed under the auspices of the IOM Committee on Aerospace Medicine and Medicine in Extreme Environments will examine NASA's Longitudinal Study of Astronaut Health (LSAH) and make appropriate medical, scientific, and administrative recommendations for improving the study, as well as recommendations relative to the data trends identified to date, inclusion of astronauts from NASA's international partners, appropriate follow-up of findings, and medical care of current and former astronauts, mission specialists, and other space travelers. In so doing the committee will address the potential relevance of lessons learned from historical exposures such as agent orange, radiation among veterans, and industrial beryllium

to the configuration of the LSAH with regard to its usefulness in identifying health risks.

GOALS AND DESIGN OF THE CURRENT LSAH

According to the most recent published description of the LSAH, the primary aim of the LSAH is:

to investigate and describe the incidence of acute and chronic morbidity and mortality of astronauts and to determine whether the unique occupational exposures encountered by astronauts are associated with increased risks of morbidity or mortality. Specifically, the primary a priori hypotheses being tested are:

1) Astronauts are at different risk of total and cause-specific mortality than are ground-based employees; and

2) Astronauts are at different risk of total and specific morbidity than are ground-based employees (Hamm et al, 2000).

Study Participants

The primary focus of the study is the group of men and women who have been selected as NASA astronauts since the space program began in 1959. This includes both pilots and mission specialists, who have been career astronauts, but not the 27 American payload specialists who generally are scientists or engineers who fly only a single mission and return to their preflight career immediately afterward (five have flown on two missions, and one on three missions). All active astronauts participate in the study. Astronauts who have retired or otherwise left NASA are invited to continue in the study, and their participation rate is high but not close to universal (the rate of return to JSC for annual exams varied from 61 percent to 88 percent over the nine years between 1993 and 2001). All the ex-astronauts are still "in the study," although some subset of them misses their exams each year. It is the opinion of the LSAH staff that the same individuals generally return every year.

The study also collects health and medical data from a non-astronaut comparison group of JSC employees. The astronauts are a highly screened group selected for specific expertise, education, and personal traits. They must also meet stringent medical standards (that were not written down until 1977 and continue to evolve), presenting a considerable challenge in constructing a useful comparison group. Military pilots, astronaut applicants who passed the medical examination criteria but were not selected, scientists who wintered over in Antarctica, and other populations were considered as possible comparison groups.

However, after careful review, it was determined that JSC civil service employees best met the need for a comparison population for this study. JSC employees have similar general ground-based occupational and background environmental exposures as do the astronauts and receive routine physical examinations in the same clinic system that conducts the physical examinations for the astronauts. Although separate staffs of physicians and nurses serve the two groups, the same technicians examine both groups using the same equipment and laboratories.

The number of astronauts selected is determined by the needs of the space program; so the number of comparison participants was selected to provide the best combination of statistical power and efficient use of resources. This was determined to be three comparison participants for every astronaut.

After each class of astronauts is selected, gender specific means and standard deviations are calculated for age and body mass index (BMI). Male and female employees who have received a physical examination at the Occupational Health Clinic within the previous three years are then identified as potential comparison participants if their age and BMI both fall within two standard deviations of the astronaut means by gender. Individuals are randomly selected from this group and asked to participate in the study. No monetary incentives are offered for participation, and the informed consent process includes statements assuring participants that there would be no adverse consequences for declining to participate or for withdrawing from the study at any time.

Selection of comparison participants is now done in the same year that their matching astronaut class is selected, but comparison participants to match astronauts selected prior to the start of the LSAH in 1992 were necessarily selected retrospectively. The first astronaut class, selected in 1959, did not train at JSC, and civilian employees at JSC were too few for a 3:1 match until 1967, so comparison participants for the astronauts selected between 1959 and 1967 were selected from JSC employees of 1967. After 1967, employee records for the year of each astronaut class were used to select comparison participants. In January 2003 the astronaut group totaled 312 and the comparison group 928.

Medical Data Collected

The primary data for the LSAH are obtained from medical records maintained at the JSC clinics. Annual health evaluations are required of active astronauts and are offered to inactive astronauts. These evaluations consist of a medical history, physical examination, laboratory tests, medical images, and other diagnostic tests. All other JSC employees were offered similar evaluations annually prior to 1994, but now they are offered to them only every three years.

Non-astronaut employees who are participating in this study are offered evaluations every other year. Details of these evaluations, which are referred to as "physical exams" in the remainder of the report, are contained below in Table 1-1. Reports and documentation of interim medical care are obtained as part of the evaluation in order to document relevant morbidity information on the examination form. Other study data are obtained from interim visits to the JSC clinics for sick calls, reports from consultants and private physicians, and hospital discharge reports (Hamm et al., 2000; Wear, 2003).

Additional data are obtained for astronauts from preflight and postflight physical examinations, medical debriefings following flights, inflight experimental data, and reports of inflight medical events. A questionnaire designed to capture lifestyle factors and health risk data was developed and first mailed to all LSAH study participants in 1994. This questionnaire is now mailed to all new participants when they enter the study. Follow-up questionnaires are mailed to each participant every two years to capture changes and new information. In addition, in 1998, usual nutrient intake was assessed with a mailed food frequency questionnaire (Hamm et al., 2000).

Mortality and cause of death are confirmed by death certificate. Biannual searches are done for death certificates of those participants who miss a scheduled examination and cannot be contacted by mail or telephone. Copies of autopsy reports and hospital death summaries to support death certificate data are obtained whenever they are available (Wear, 2003).

The actual variables collected and the frequency at which they are collected have changed since the study began, primarily as a result of budgetary constraints. Table 1 shows the measures in the database and the collection schedules for both the astronaut and comparison groups as of January 2003. Appendix B provides a fuller description of each of the measures in the table.

TABLE 1-1 Physical and Health Measures Collection Schedule - 2003 LSAH

Measures	Astronauts	Comparisons	Notes
Physical exam	Annually	Every 2 years	
Dental exam	Annually, but for active duty only	Never	
Visual acuity	Annually	Every 2 years	
Color vision	Annually	Every 2 years	
Depth perception	Annually	Every 2 years	
Heterophorias	Annually	Every 2 years	
Intraocular pressure	Annually	Every 2 years	
Audiometry	Annually	Every 2 years	
Electrocardiogram (ECG)	Annually	Every 2 years	
Pulmonary function by standard spirometry	Annually	Every 2 years	
DEXA scan*	Every 3 years	Never	
Exercise tolerance test (85% max)	Age-specific intervals (US Preventive Services Task Force Guidelines)	Age-specific intervals (US Preventive Services Task Force Guidelines)	51+ = annually Comparisons every 2 years
Colonoscopy	Age 40,50,60,70,80	Age 40,50,60,70,80	
Proctosigmoidoscopy	Age 45,55,65,75	Age 45,55,65,75	
Mammogram	Age-specific intervals (US Preventive Services Task Force Guidelines)	Age-specific intervals (US Preventive Services Task Force Guidelines)	50+ = annually. Comparisons every 2 years
Pelvic exam	Annually	Every 2 years	
Pap smear	Annually	Every 2 years	
Hematology	Annually	Annually	
Lipid profile	Annually	Annually	
Urinalysis	Annually	Annually	
Chemistry panel	Annually	Annually	
Prostate specific antigen	Annually after age 40	Annually after age 40	
Immunoglobin panel	Annually	Every 4 years	
Serology	Annually	Every 4 years	
Serum protein panel	Annually	Every 4 years	
Self-reported medical history	Annually	Annually	

TABLE 1-1 Continued

Measures	Astronauts	Comparisons	Notes
Medical records from JSC clinics, private physicians, and hospitals	Annually, as follow-up to physical exam	Every 2 years, as follow-up to physical exam	
Lifestyle questionnaire	Every 2 years	Every 2 years	
Death records	Upon notification of death by any source, information I independently verified.	Upon notification of death by any source, information is independently verified.	Stimulus for professional search is missed exam and no reply to calls or letter
Postflight medical debrief	Postflight	Not applicable	

* DEXA: dual energy x-ray absorptiometry.
SOURCE: Wear, 2003.

DATA ACCESS POLICY

A detailed protocol describing the policies and procedures involved in accessing the LSAH data has been elaborated and published as Section 6 of The LSAH Manual of Procedures. The following information from that document provides an overview of the procedures in place.

The purpose of the LSAH Executive Committee is to ensure that data quality is maintained, the variables are interpreted consistently, there are no redundant projects, and the confidentiality of the participants' medical data is maintained. To meet its objectives, the Executive Committee

- reviews and approves all requests for LSAH data before releasing any data from the study database
- reviews and approves any presentation or publication of the data
- maintains a permanent file of all requests and the subsequent actions regarding each request

The Executive Committee consists of the

- NASA LSAH Technical Monitor
- Chief, Flight Medicine Clinic

- Chief, Medical Operations Branch
- Assistant to the Director, Space Medicine
- Assistant to the Director for Biomedical Research and Countermeasures
- Section Supervisor, Epidemiology

Extramural requests for data must receive initial merit and funding approval via the National Space Biomedicine Research Institute (NSBRI) or a NASA Research Announcement (NRA) before they are submitted to the LSAH Executive Committee. If an extramural investigator requests extensive data retrieval and analyses, the Executive Committee may require that the investigator provide the necessary funding to support this work.

Intramural data requests are categorized into those for research, clinical care, or operational purposes. Research questions may be submitted by JSC civil servants in the Space Life Sciences Directorate. Contractors, residents, and postdoctoral students and fellows must obtain a civil servant sponsor before submitting their data requests to the Executive Committee. These requests must obtain approval through an independent peer review process before being submitted to the LSAH Executive Committee. Until this process is officially in place, the LSAH Executive Committee will serve this function.

Clinical care questions focus on direct patient care and are submitted by JSC Flight Surgeons. These data requests are submitted to the NASA Technical Monitor or the Epidemiology Section Supervisor. Approval by the Executive Committee is not required because the purpose is to support clinical care of individual patients. However, if the Flight Surgeon later wishes to publish or present the results, a study protocol must be developed and submitted to the Executive Committee for review and approval.

Data requests for operational or management purposes do not require Executive Committee approval, but results later determined to be publishable require a protocol before they are submitted to the Executive Committee for review and approval.

2

Findings to Date

Because the Longitudinal Study of Astronaut Health (LSAH) began in 1992, more than three decades after the first astronauts were selected, the study's database was rapidly populated with a wealth of retrospective data. As a result, it was possible to query the data early in the study's course. The first peer-reviewed publication appeared in 1993 (Peterson et al., 1993). It reported astronaut mortality from 1959 through 1991. This chapter summarizes the findings of that publication, two other peer-reviewed papers from later in the 1990s, and more recent analyses of morbidity and mortality that National Aeronautics and Space Administration (NASA) scientists provided to the committee in meetings held in 2003.

Hamm et al. (Hamm, 2000) updated Peterson's figures but followed the same sample for seven more years. During that time, six members of the 295 astronaut sample died. Therefore, the numbers will vary according to which paper is quoted.

Although some published reports utilizing the LSAH are included, the committee concentrated on the organization, goals, and function of the LSAH rather than a critique of the methods and analyses, which has already passed peer review.

PETERSON ET AL., 1993

The space radiation environment was a major concern for NASA from the earliest days of the space program. Space travelers are exposed to radiation that is different from that to which terrestrial workers, such as those in the nuclear power or nuclear weapons industries, are subjected. A National Academy of Sciences panel was asked to develop radiation protection guidelines and identify biological responses for human exposure to space radiation (National Academy of Sciences, 1967), and NASA has maintained a database on space (and medical) radiation exposure for all astronauts since Project Mercury began in 1959. It is not surprising that the initial analysis of the LSAH data (Peterson et al., 1993) focused not simply on mortality but also addressed the hypothesis that astronauts are at increased risk for malignant neoplasms.

The study looked at the medical records of the 195 astronauts selected between 1959 and 1991 and found that 20 deaths had occurred during the 32 years surveyed. Sixteen were due to spacecraft (8), aircraft (7), or automobile (1) accidents; 2 were due to circulatory disease; 1 was the result of a malignant neoplasm; and 1 was due to unknown causes. Standardized mortality ratios (SMR) based on the U.S. population, adjusted for age, race, gender, and calendar year, were significantly increased for all-cause deaths (SMR=181) and accidental deaths (SMR=1,346). The crude accidental death rate of 445 deaths per 100,000 person-years for the 12 occupationally related deaths was an order of magnitude greater than the 34 to 41 per 100,000 typical of the mining industry, although the SMR for all accidents was comparable to that reported for Canadian airline pilots in another study (Band, et al., 1990). The hypothesis that astronauts are at increased risk for cancer mortality compared to the U.S. population was not supported, although the relatively young age of the astronauts, the low doses of radiation during space flight, the modest interval between space flight and data analysis, and the small sample size all made statistical confirmation unlikely. Space radiation doses varied directly with mission duration (r=0.99), and average mission doses ranged from less than 0.1 milliGray (mGy) for Mercury astronauts (average mission duration of less than 1 day) to 43 mGy for Skylab astronauts (average mission duration of 57 days). The average dose for astronauts on the first 43 shuttle missions was 1.3 mGy. For each of the 13 astronaut classes examined, the average per capita dose of radiation from diagnostic medical X-rays exceeded that from space travel—in most of the earlier years by a factor of 10 or 12.

HAMM ET AL., 1998

This study (Hamm et al., 1998) focused on cancer mortality in a slightly larger population of astronauts than that of the Peterson et al. study of 1993, and it included an LSAH comparison group of JSC civil servants matched to the astronaut group on age, sex, and body mass index (BMI). The observed rate of cancer mortality in each of those groups was also compared with the age- and sex-adjusted cancer mortality rates of residents of Public Health Region 6 of Texas. All 3 cancer deaths among astronauts through 1995 were in males, so the investigators chose to confine their analyses to the 210 male astronauts selected through that date. The 618 civil servant comparison participants constituted all the males in that group, and the data from the Texas general populations included only males as well.

Both the astronaut group and the Johnson Space Center (JSC) comparison group had lost three members to cancer. The cancer mortality rate of the astronauts appeared increased compared to that of the comparison group (SMR =

345; 95 percent confidence interval (CI) = 66 to 756), but the apparent increase was not statistically significant. Both groups showed much lower than expected rates of death when compared to residents of Texas Public Health Region 6 (SMR for astronauts = 47; CI = 10 to 105; SMR for comparison group = 17; CI = 4 to 38). For the comparison group the difference was statistically significant. The cancer types causing the astronaut deaths were undifferentiated carcinoma of the nasopharynx, glioma, and metastatic melanoma. The comparison group fatalities were due to metastatic melanoma (2) and glioblastoma multiforme. The lack of a significant difference between the astronaut and control groups is again not unexpected, given the small number of cases, the short duration of the astronauts' space experience (mean of 12.6 days), the low dose of space radiation (mean of 1.65 mGy), which is not significantly different from the background radiation, and the relatively young age of both groups. Additionally, the cancers found are not clearly linked to ionizing radiation. The comparisons to cancer rates in the general population are also not too surprising, given the substantially higher levels of education, income, general health, and fitness that characterize the astronaut group and their JSC comparisons. Employment itself is well known to be associated with lower mortality rates than those of the general population—"the healthy worker effect" (Fox and Collier, 1976).

The statistical analyses were confined to cancer mortality, but the report alludes to a preliminary review of the medical records that indicated that there had been at least 21 nonfatal cancer cases among the astronaut group and at least 6 cases in the comparison group. Non-melanoma skin cancers accounted for 17 of the 21 astronaut cases and 3 of the 6 comparison cases.

HAMM ET AL., 2000

This study, published by Hamm et al. in *Aviation, Space, and Environmental Medicine* in 2000, updated mortality data still further, but it was primarily devoted to describing the study design and baseline data from the initial health evaluations of all the participants, (i.e., both comparisons and astronauts), who entered the study between 1959 and 1991. The baseline data—demographic, behavioral, and physiologic—are important indications of how closely the comparison participants might match the astronauts in initial propensity for disease.

Not surprisingly, the ages and body mass index of the comparison participants closely approximated those of their astronaut counterparts (see Table 2-1). Caucasians comprised 94 percent of the astronaut group and 90 percent of the comparison group. Women comprised approximately 11 percent of each group. Other demographic data reported were marital status (84 percent of each group were married) and education. All astronauts have at least a bachelor's degree at selection, and 77 percent of those had a graduate degree as well. Only 36 percent of the comparison participants had an advanced degree ($p = 0.001$) and 6.6 percent had less than a bachelor's degree.

TABLE 2-1 Mean (Standard Deviation) Age in Years and Body Mass Index (BMI) at Selection of LSAH Participants, 1959-1991

Males				Females			
Astronauts (n=175)		Comparisons (n=510)		Astronauts (n=20)		Comparisons (n=65)	
Age	BMI	Age	BMI	Age	BMI	Age	BMI
33.3 (3.0)	23.6 (1.9)	32.9 (4.3)	23.5 (1.5)	30.9 (2.6)	20.8 (2.2)	30.6 (2.3)	21.0 (2.7)

SOURCE: Hamm et al., 2000.

T-tests were used to compare the astronaut and comparison groups on the substantial number of measures derived from baseline physical examinations and clinical laboratory tests (see Appendix B for a full list). Astronauts had significantly lower pulse rates, systolic and diastolic blood pressure, hemoglobin, and serum triglycerides, and significantly higher blood glucose. Seventy-nine percent of the astronauts had uncorrected visual acuity of 20/20 or better, and none had worse than 20/150, while only 55 percent of comparisons had 20/20 vision and 23 percent of comparisons had acuity worse than 20/150. No other statistically significant differences were identified.

The report also notes that eight comparison participants (1.6 percent) had controlled hypertension, and one had borderline hypertension at selection. Two of the comparison participants had diabetes at selection. These are disqualifying conditions for astronaut selection, so there were no cases of either hypertension or diabetes in the astronaut group at selection.

Twenty-six of the astronauts and 14 of the comparison participants had died at the time the Hamm et al. (2000) report was written. Table 2-2 shows the causes as well as relative risks.

Just as the Peterson (1993) report found, the only cause of death found to be significantly different between the two groups was injury and accidental deaths. The astronauts are clearly at a greater risk of accidental death than are the comparison participants. Eight of the accidental deaths among the astronaut group were due to two spacecraft accidents. Five deaths among this group were due to crashes of high-performance military aircraft, and three were due to accidents involving commercial or private aircraft. One astronaut died of altitude sickness and exposure to cold. One astronaut and one comparison died in car crashes. The other accidental death among the comparison population was due to a firearm.

Cancer mortality still appeared greater in the astronauts, but the apparent difference between the groups was not statistically significant.

TABLE 2-2 Cause-Specific Mortality among Longitudinal Study of Astronaut Health Participants Selected from 1959 to 1991

	Astronauts (N=195) (Person-years=3,901)		Comparisons (N=575) (Person-years=12,471)					
Cause of Death	Deceased	Percent	Deceased	Percent	Crude RR	Adjusted RR*	95% CI	*p* Value
Cancer	4	2.05	3	0.52	4.26	3.19	0.93-21.85	0.2382
Cardio-vascular	3	1.54	7	1.22	1.37	1.20	0.27-5.28	0.8112
Acci-dents and injuries	18	9.23	2	0.35	28.77	22.91	5.02-104.46	0.0001
Other diseases	1	0.51	2	0.35	1.60	2.27	0.21-25.22	0.5040
Total	26	13.33	14	2.43	5.93	5.07	2.46-10.41	0.0001

*Adjusted RR (relative risk) was adjusted for sex, education, marital status at selection, and smoking history using proportional hazards regression. Missing values made it impossible to adjust for physiological measures.
Confidence intervals (CI) and *p* values are for the adjusted relative risk.
SOURCE: Hamm et al., 2000.

BRIEFINGS OF THE IOM COMMITTEE

In January 2003 and again in March 2003, NASA scientists from the Space and Life Sciences Directorate at Johnson Space Center briefed the Institute of Medicine (IOM) Committee on its analysis of LSAH data concerning overall mortality and the three clinical conditions that the committee was asked to review (cataracts, cancer, and thyroid function).

Mortality

James Logan summarized the LSAH data on mortality from all causes for the committee (Logan, 2003). As in the earlier reports summarized above, overall mortality has been significantly higher for the astronaut group. Logan's presentation in January 2003, just prior to the loss of the space shuttle Columbia and its crew of 7, reported 29 deaths among the 312 astronauts in the LSAH database and only 17 deaths among the 912 comparison participants. Accidental deaths, including 8 in spacecraft losses (3 in the *Apollo* fire and 5 in the *Challenger* explosion), accounted for 20 of the astronaut deaths (versus only 2 in the comparison group).[1] The groups did not differ significantly in mortality from any other cause.

Men accounted for 27 of the 29 (93 percent) of astronaut deaths and all 17 of the comparison group deaths. The 12 astronaut accidental deaths that were not spacecraft-related included 4 in T-38 jet trainer crashes, 4 in private plane crashes, 1 in a commercial plane crash, 1 each in car and motorcycle crashes, and 1 while mountain climbing.

Box 1-1 Radiation Terms and Measurement Units

Absorbed dose is the energy actually deposited in a certain mass of tissue. It does not take into account either the differing biological effects of the different radiation types or the differing responses of different tissue types. The international unit (SI) is the gray (Gy), which is equivalent to the absorption of 1 Joule of energy per kilogram of mass. An older unit is the rad. One Gy equals 100 rad.

Equivalent dose accounts for the different effects the various types of radiation have on biological tissue. It is calculated by multiplying the absorbed dose by a radiation-specific weighting factor (w_R) or quality factor determined by the International Commission on Radiological Protection (ICRP). The SI unit of equivalent dose is the sievert (Sv); the older unit is the rem. One Sv equals 100 rem.

Effective dose accounts for the varying sensitivity to radiation of different tissue types (skin, bone, brain, etc). It is a composite whole body dose calculated by multiplying each tissue type by an ICRP tissue weighting factor (w_T) and summing the weighted equivalent doses. This composite dose is proportional to the increased risk from cancer and genetic effects. The SI unit of effective dose is Sv.

Cataracts

Frank Cucinotta reported to the IOM committee (Cucinotta, 2003) that an optometrist who had performed annual eye examinations of astronauts for more than a decade told the LSAH staff in 1998 that he had seen numerous lens opacities among the astronauts, possibly even more frequently than in his private practice. The LSAH staff immediately initiated an investigation by the medical staff and radiation group, which resulted in the publication of a detailed analysis

[1] There were seven deaths in the *Challenger* accident, but only the five astronauts are included in the LSAH. The other two casualties were payload specialists not included in the LSAH.

by Cucinotta et al. (2001), followed by recommendations to reduce radiation exposure (see Box 1-1).

Astronauts have had eye examinations since the beginning of NASA, but at varying frequency during the early years. Prior to 1977, examinations were performed at the JSC flight clinic. Between 1977 and 1988 they were done by referral physicians in the Houston area, and the results were collected by JSC. Since 1989 they have again been performed at the JSC flight clinic. Forty-eight lens opacities have been found in 295 astronauts. No measurements of astronaut exposure to ultraviolet radiation have been made, but each astronaut except for those on the initial four Mercury flights has worn a thermoluminescent dosimeter badge while in space.

Cucinotta and his colleagues used those badge data to calculate a lens equivalent dose for space radiation. They then developed an extensive database on the exposure of these astronauts to radiation from both diagnostic medical X-rays and occupational aviation. They sorted the astronauts into high-dose and low-dose groups and computed relative hazard ratios for cataracts at age 60 and at age 65.

Table 2-3 shows a significant increase in cataract risk for astronauts in the high space lens dose group for all cataracts and nontrace cataracts. Hazard ratios using lens dose from medical X-rays alone and from aviation alone were not significant. There was a significant association between cataracts and high-inclination or lunar missions, where a much higher flux of heavy ion radiation occurs. Ninety percent of the 39 cataracts occurring after space flight were in astronauts on such missions.

It has long been suspected that exposure to solar particle events, galactic cosmic rays, and trapped protons and electrons would increase the lifelong risk of developing cataracts. Early studies of cancer patients suggested a dose threshold for cataracts of about 2 Gray, but these data from Cucinotta et al. (2001) showed that astronauts with exposures above 8 milliSieverts developed cataracts more frequently and at an earlier age than those exposed to less than 8mSv (Table 2-3). Space radiation has higher linear energy transfer than terrestrial radiation, and this study reported exposure in tissue penetration rather than surface measurement (so the absorbed dose (Gy) and the equivalent dose (mSv) would be approximately equal).

NASA has contracted with Dr. Leo Chylack, an ophthalmologist at Harvard, for a five-year follow-on study comparing the prevalence and rate of progression of cataracts in astronauts with those in a group of current and former military pilots matched to the astronauts for age and gender. Digital photography and computerized image analysis will ensure comparable, objective, and quantitative measurements for all subjects (Cucinotta, 2003).

TABLE 2-3 Relative Hazard Ratios and (95% Confidence Intervals) Comparing High Exposure-Group to Low Exposure Astronaut Groups for Cataract Risk at Age 60 and at Age 65.

Cataract Type	Ratios using lens dose from all radiation sources*	Ratios using lens dose from space radiation only**
At Age 60 Years		
All cataracts	1.51 [0.64, 3.59]	**2.35 [1.01, 5.51]**
Nontrace cataracts	2.47 [0.76, 8.01]	**8.04 [2.51, 25.7]**
At Age 65 Years		
All cataracts	1.88 [0.93, 3.83]	**2.44 [1.20, 4.98]**
Nontrace cataracts	**3.85 [1.45, 10.2]**	**7.26 [2.74, 19.3]**

*Relative hazard of astronauts with total lens dose >35millisieverts (mSv)(average 70 mSv) compared to those with lens dose <35 mSv (average 20 mSv). Statistically significant values are in bold type.

**Relative hazard ratio of astronauts with a space lens dose >8 mSv (average 45 mSv) compared to those with lens dose < 8 mSv (average 3.6 mSv).). Statistically significant values are in bold type.

SOURCE: Cucinotta et al., 2001.

The Committee considers this work an excellent example of the potential value of the LSAH, but it notes that the impetus for the study was the anecdotal report of more frequent cataracts by an examining doctor with no direct tie to the LSAH. When this suspicion was reported, 48 cases of lens opacification were subsequently culled from the data for 295 astronauts, and the problem was referred to the radiation and space groups at JSC for more detailed study.

Cancer

Because of the known association of some cancers with radiation exposure, surveillance of astronauts for malignancies was planned from the beginning of the LSAH. Craig Fischer briefed the committee on the comparison of cancer incidence among the astronauts (Fischer, 2003), the LSAH comparison participants, and an age- and sex-matched sample of the National Cancer Institute's Surveillance, Epidemiology, and End Results (SEER) database. Fourteen cases of cancer were diagnosed among the 312 astronauts followed from 1959 to the present. This is 59 percent higher than the comparison group per person/year (not statistically significant), but 46 percent lower per person/year than the SEER data (statistically significant). The distribution of the astronaut malignancy types is shown in Table 2-4. The prostate is the predominant cancer site in both the astronaut and comparison groups.

TABLE 2-4 Number and Type of Cancers Diagnosed in NASA Astronauts and LSAH Comparison Group Participants

Diagnoses	Number
NASA astronauts (N=312)[a]	
Nasopharyngeal carcinoma	1
Malignant melanoma, skin	1
Malignant melanoma, primary site unknown	1
Renal cell carcinoma	1
Papillary carcinoma, thyroid gland	1
Hodgkin's disease, NOS	1
Leukemia, NOS	1
Lymphoma, primary in brain	1
Adenocarcinoma, prostate	4
Carcinoma, gall bladder	1
DCIS & LCIS, breast, bilateral	1
Total	14
LSAH comparison participants (N=928)[b]	
Malignant melanoma, skin	6
Adenocarcinoma, prostate	12
Papillary carcinoma, bladder	2
Malignant neoplasm, testis, NOS	1
Malignant neoplasm, cervix, NOS	1
Malignant neoplasm, larynx	1
Adenocarcinoma, colon	1
Malignant neoplasm, brain, NOS	2
Total	26

[a]Excluded are 33 diagnoses of basal and localized squamous cell carcinomas of skin.
[b]Excluded are 27 diagnoses of basal and localized squamous cell carcinomas of skin.
NOS, not otherwise specified.
DCIS, disseminated carcinoma in situ; LCIS, localized carcinoma in situ
SOURCE: Fischer, 2003.

Thirty-three cases of basal and squamous cell carcinomas of the skin were excluded from the analysis of the astronaut group, and 27 were excluded from the 912 member comparison sample, an almost threefold difference in rate. A rationale offered for this deletion is that astronauts spend significant time outdoors for both training and recreation, but this is without supporting data and is inconsistent with the professed goal of assessing the overall risks of being an astronaut. The higher than threefold increase in prevalence in astronauts is statistically significant. Including these non-melanoma skin cancers with all other cancers would make the overall difference between the astronaut group and comparison group significant as well.

Thyroid Function

Kathleen McMonigal briefed the IOM committee on NASA's discovery of the adverse effects of iodinization of space shuttle drinking water (McMonigal, 2003; McMonigal et al., 2000). Although the role of the LSAH in detecting these thyroid function abnormalities was indirect, it is relevant to the discussion of LSAH design and execution.

In 1990 a female astronaut was diagnosed as hypothyroid by her flight surgeon. In the course of their discussion, the surgeon noted that he had seen several such cases in astronauts, but they were young men, in which hypothyroidism is rather unusual. Daily thyroid hormone replacement therapy was prescribed for the astronaut, and she subsequently flew two more shuttle missions without difficulty. The thyroid problem resurfaced in 1997 when Dr. McMonigal was monitoring a terrestrial project in which life support systems were being tested by 4 astronauts in isolation for 30, 60, and 91 days. She was told that there might be a problem with the iodine concentrations in the water and was asked to check the thyroid function of one of the astronauts in who had participated in the previous 60-day test. He had started with a high normal thyroid stimulating hormone (TSH) level and subsequently became hypothyroid during the test.

The physician discovered that the iodine levels in the test subjects' drinking water was 5 milligrams per liter and the concentration of iodine in the water on the previous shuttle missions had ranged from 3 to 4 mg per liter. The decision was made to test the new group of 4 euthyroid astronauts at 30 days, and if thyroid function abnormalities were found, reduce the iodine concentration. The recommended daily allowance (RDA) for iodine is about 0.15 mg, but these subjects were ingesting more than 10 to 20 mg in drinking water alone. The danger of ingesting such doses has been known for many years. The Wolff-Chaikoff effect (high doses of iodine block the organification of thyroglobulin) was described more than half a century ago (Wolff et al., 1949), and the Jod-Basedow effect (potential hyperthyroidism after iodine administration) more than a century ago. All four astronauts in the test showed marked elevations of TSH. Anion exchange resin filters were then installed at the tap, which lowered the iodine concentration to 0.25 mg per liter (approximately a 16-fold reduction), and thyroid function measures returned to normal during the period of observation. Two of the eight test subjects studied showed functional abnormalities, one hypothyroid and one hyperthyroid, which returned to normal in about a year.

The episode described above prompted a review of the LSAH database for evidence of thyroid abnormalities in the astronaut population, because high concentrations of iodine had been the bacteriocide of choice on U.S. space vehicles to that point. Analysis of postflight TSH levels indicated that a transient but statistically significant elevation over preflight levels on the day of return was common. By the next annual physical examination, TSH values had typically returned to preflight levels. Mean preflight TSH level for all 134 astronauts for whom data were available at the time (1998) was 2.84 microInternational units per ml (μIU/ml) (SD=2.24). Upon return, mean TSH was 3.43 μIU/ml (SD=2.59). SKYLAB astronauts (n=9) remained in orbit an average of 57 days, and they had much higher TSH levels immediately upon return (8.44+/– 6.94) and three days after return (8.56 +/- 3.70), although the levels eventually returned to preflight levels (4.89 +/- 3..65). U.S. astronauts who flew on the Russian space station MIR for periods of 2-6 months (n=6) showed no significant increase in TSH upon return. Mir's drinking water was treated with silver nitrate rather than iodine. Postflight elevations in TSH are also not seen in data from the 79 astronauts who have flown on the space shuttle after iodine levels in the drinking water were reduced to 0.25 mg/liter in 1998.

The LSAH data were also searched for eight ICD codes related to clinical thyroid disease. Thirty-nine cases were uncovered by this 1998 search. Nine male astronauts (3.8 percent of all male astronauts) and 3 female astronauts (8.3 percent of all female astronauts) were found to have clinical thyroid disease, along with 18 male (2.5 percent) and 9 female (7.8 percent) comparison subjects. The overall odds ratio comparing astronaut incidence to comparison group incidence was 1.39 (95 percent CI=0.69 to 2.78). For males only, the odds ratio was 1.56 (CI=0.67 to 3.63). For females only, the odds ratio was 1.07 (CI = 0.27 to 4.18). None of these comparisons was statistically significant, nor were further tests for association of disease with estimated iodine intake during mission, gender, or role (pilot or mission specialist).

3

Issues with Design and Implementation of the Current Longitudinal Study

In an ideal world, the Longitudinal Study of Astronaut Health (LSAH) would provide the answer to a cascading series of questions about long term risks among three populations (see Table 3-1).

TABLE 3-1 LSAH Research Questions and Appropriate Study Populations

Research question	Focus population
1. What are the long-term risks of space-flight?	Astronauts with history of spaceflight.
2. What are the long-term risks of preparing for spaceflight?	Above population PLUS astronauts who have never flown in space.
3. What are the long-term risks of working in the environment of JSC?	Above populations PLUS employees of JSC.

To get an unbiased estimate of these cascading risks, a series of studies with different populations and comparison groups might be designed and carried out. In each of these the astronauts and their comparison group controls should

- be equivalent at baseline in all factors that influence risk of disease or adverse health outcomes;
- have equivalent exposures in day-to-day life except for those related to exposure to spaceflight or preparations for spaceflight;
- have equivalent monitoring for disease by observers blinded to whether they were exposed to spaceflight or preparations for spaceflight; and
- participate fully from study entry to the outcome of interest.

In the real world, like many expensive, long-running epidemiological studies, the LSAH has had to make a number of compromises. The remainder of this chapter systematically describes and examines those compromises and their effects on the utility of the LSAH.

GOALS OF THE LSAH

The published goals of the LSAH appear straightforward at first glance. Hamm et al. (2000) list them as follows:

> The primary aim of the current LSAH is to investigate and describe the incidence of acute and chronic morbidity and mortality of astronauts and to determine whether the unique occupational exposures astronauts encounter are associated with increased risks of morbidity or mortality. Specifically, the primary a priori hypotheses being tested are:
>
> - Astronauts are at different risk of total and cause-specific mortality than are ground-based employees; and
> - Astronauts are at different risk of total and specific morbidity than are ground-based employees.

The Manual of Procedures for the LSAH explains the purpose of the study in several places. The first lines of Chapter 1 state that the purpose is to "examine the incidence of acute and chronic morbidity and mortality of astronauts and describe the risks of morbidity and mortality associated with the astronauts' occupational exposures, as compared with the risks for civil service employees of Johnson Space Center." It goes on to say that assessments of lifetime risk and flight-time risks for specific diseases and disorders "will help to (1) identify health-related problems that require spacecraft medical facilities, and (2) devise and implement methods to reduce risks." Chapter 2 of the Manual includes a copy of the information handout given to potential comparison participants. Under "goals" it includes the statement: "These assessments will be useful in identifying potential health-related problems resulting from exposure to spaceflight." The letter of invitation sent to potential comparison participants is even more specific, noting that the LSAH has been established "to evaluate the long-range medical effects from exposure to microgravity."

One study could conceivably encompass all three goals with variations in data analysis, but the committee believes that whatever the goal, two different uses of the data are required, each of which imposes slightly varying demands in design and execution of the study. The first of these potential uses is to provide scientific documentation of the long term effects of space flight and preparation for space flight, and thus serve as an indicator of necessary or desirable modifications to current and future training, spacecraft, or operating procedures for future flights that would enhance the safety of spaceflight for future astronauts. To be useful in this regard, the data would need to be queried regularly but especially when changes in the space program are contemplated (e.g., transition from short-duration shuttle missions to much longer duration missions on the International Space Station, or from the shuttle to a second-generation reusable launch vehicle).

The second potential use of the LSAH is as a surveillance instrument to facilitate rapid prevention, detection, and treatment of occupation-related health problems in the group of current and former astronauts. To accomplish this mission, data entry would have to be far more rapid than it has been, and analysis far more frequent. The committee believes that insofar as participation in the LSAH is mandatory for active duty astronauts, NASA is ethically bound to include this surveillance mission along with the research mission. In practice that means that not only should test results and other measures be passed on to individual astronauts and their physicians for individualized evaluation and clinical evaluation, but any group-related information that could influence individual health care must be fully and quickly shared with all affected parties, including payload specialists and astronauts from other countries.

The two missions described present different and sometimes conflicting challenges, and striking a balance is no mean feat, but it appears to the committee that, at this point, the LSAH is serving only the first of the two purposes. A case could be made that LSAH data on thyroid function were useful in the decision to reduce the amount of bactericidal iodine used in the shuttle's potable water, but as noted in the previous chapter, the LSAH seems to have played only a confirmatory role in the discovery of iodine-induced thyroid dysfunction in shuttle astronauts. There is certainly no evidence that the LSAH has ever been queried regularly enough and with sufficiently powerful pattern detection methods to serve as the basis for an effective prevention program for current and former astronauts. That is not to minimize the salutary effects of the annual physicals and other testing of astronauts and former astronauts on the health of individual participants, but those benefits have resulted primarily from individual history and physical and laboratory examinations rather than the study findings. An exemplary surveillance effort requires more frequent reviews and analyses of the accumulating data than a pure research project, and inclusion of additional data as well. No mental health data are included in the current LSAH dataset, for example, although there is no reason to believe that astronauts are immune from psychiatric problems. Such data might also prove useful in the research aspect of the LSAH as well, because many of the other measures that are being collected in this study can be influenced by mental disorders and personality variables. For example, the presence of an eating disorder can have a dramatic impact on measures of bone density. Depression is also increasingly recognized as having a strong negative effect on the morbidity and mortality of many other comorbid conditions if not recognized and treated promptly.

A second difficulty related to the goals and purposes of the LSAH is whether the LSAH is primarily a study of the health consequences of being an astronaut or a study of the long term effects of space flight. Analyses of the data by NASA to date have been consistent with the declared purpose of assessing the occupational risks of a career as an astronaut, but they have not always been suited to the goal of "identifying potential health-related problems resulting from exposure to spaceflight" (*Manual of Procedures*, Section 2.2). The recent analysis of cancer prevalence (Fischer, 2003), for example, includes astronauts who had not as yet flown a space mission, even though risk for cancer due to exposure to the unique radiation environment of space has been a major concern since the beginning of the space program (Peterson et al., 1993). The database is clearly sufficient to investigate both hypotheses (i.e., that members of the astronaut corps are at no greater risk for cancer than Johnson Space Center [JSC] civil servants, and that spaceflight does not increase the risk of cancer), but the analysis required would be different in each case. If the long-term risks of spaceflight are truly a concern, not only should analyses exclude astronauts who have yet to fly a mission, but some additional power would be gained by including the 27 payload specialists who have flown in space but are not included in the LSAH. Some of these specialists have logged more hours in space than many career astronauts.

STATISTICAL POWER

An important question for the design of any study is whether it has sufficient power to detect differences among the subject groups in the variables of interest. In the case of the LSAH, this means asking how likely we are to detect a difference between the astronaut and comparison groups in the prevalence of a specific health problem given that astronauts really are at higher risk for that health problem. The two primary pieces of information needed to answer that question are the prevalence of the problem and the number of subjects in the study sample. The number of subjects in the LSAH (312 astronauts, 928 comparisons) is substantial, but small for an epidemiological study and not likely to increase substantially, at least on the astronaut side. More problematic is the fact that the LSAH will potentially be queried about a large and unspecified number of health conditions with widely varying prevalence. That means that the study's power to detect group differences will vary widely as well. Table 3-2 shows in a readily understandable way how those differences limit the ability of the study to demonstrate increased risk of relatively uncommon health conditions.

The table shows the minimum detectable relative risk (for astronauts relative to the comparison participants) for health conditions varying in prevalence in the comparison group from 0.1 percent to 10 percent. Power (the probability of finding an effect given that there is one) was held constant at 0.80, and minimum detectable relative risks were computed for three different significance levels (α, the probability of falsely concluding that there is an effect when there is none). For health conditions with a 10 percent prevalence in the comparison group, for example, the astronaut group would have to have a prevalence of 14.9 percent (1.49 times the prevalence in the comparison group) to demonstrate an effect at the 0.05 level of statistical significance. Demonstrating significant differences for less common health conditions demands much larger effects. For disorders with a prevalence of only 0.1 percent in the comparison group would require a prevalence 11.15 times higher among the astronauts for the group difference to reach conventional ($\alpha = .05$) statistical significance. Under these circumstances, it will be especially important to remember that absence of evidence is not evidence of absence.

TABLE 3-2 Minimum Detectable Relative Risk (Astronauts versus Comparisons) at Different Criteria for Statistical Significance (two-tailed α) with Power = 0.80

Disease Prevalence in Comparison Group	Minimum Detectable Relative Risk		
	$\alpha = 0.05$	$\alpha = 0.10$	$\alpha = 0.15$
0.1 percent	11.15	9.27	8.18
0.5 percent	3.89	3.44	3.17
1.0 percent	2.81	2.55	2.39
2.0 percent	2.18	2.02	1.92
5.0 percent	1.70	1.61	1.55
10.0 percent	1.49	1.43	1.39

COMPOSITION OF THE COMPARISON GROUP

The procedures used for selecting comparison participants from among JSC civil servants have been successful in producing a cohort of individuals closely matched to the astronauts in gender, race, age, and body mass index. The comparison group is less well-matched in education level, but it too is a well-educated group that is much closer to the astronauts in this respect than is the general population. Like the astronauts, they were all gainfully employed at selection and have received regular preventive medical care since they were hired. Like many of the astronauts, they have spent a considerable portion of their adult lives living in Houston. The Flight Medicine Clinic and Occupational

Medicine Clinic at JSC provide physical exams and treatment of job-related injuries and illnesses for the astronauts and JSC employees, respectively. In all these respects, JSC employees make a far better comparison group for the astronauts than the general U.S. population.

On the other hand, JSC employee comparison participants may differ from the astronauts in some important unmeasured characteristics. Their physical activity and endurance may be less, and they may be less motivated to achieve and maintain excellent physical conditioning. They may perceive that the reporting of illness will be less threatening to the security of their jobs and thus be more compliant in such reporting. Other unmeasured psychological variables such as risk-taking, harm-avoidance, and competitiveness are both relevant to morbidity and mortality, and they are likely to be different in the two groups.

For many of the variables that might affect detection of, susceptibility to, or recovery from specific diseases of interest, the match between the comparison participants and the astronauts is simply unknown. For example, a longitudinal study focused exclusively on cancers would match comparisons on the basis of family cancer history, exposure to known carcinogens other than tobacco (e.g., sunlight, industrial chemicals, diagnostic and therapeutic X-rays), or genetic predisposition (e.g., BRCA1 and BRCA2). Some of this information may well be available in the LSAH database, and it may be possible to adjust statistically for it during data analysis, but the point remains that trying to create an all-purpose comparison group ironically produced one that is less than ideal for any specific health problem.

ASCERTAINMENT BIAS

It is a truism that the harder one looks for something the more likely one is to find it. Yet this is a serious flaw in the current design of the LSAH. Budgetary shortfalls over the years since the LSAH began have led to increasing disparities in the searches for health problems in the astronaut and comparison groups. The latter now receive physical exams half as frequently as the astronauts; some laboratory tests and physical evaluations that the astronauts receive yearly are provided to comparison participants only at four-year intervals; some procedures (Dual energy X-ray absorptiometry [DEXA] scans, dental exams, comprehensive visual exams) are provided only to astronauts; and there is less follow-up between routine examinations for the employees than for the active duty astronauts (see Appendix B). An additional, though perhaps minor and unavoidable, source of ascertainment bias is that the data collectors, the examining physicians in particular, are not blind to the status of the participants. The flight surgeons examining the astronauts may well look harder for health problems they believe to be associated with spaceflight, while the occupational health clinic physicians examining the comparisons pay more attention to problems that have been identified or that they suspect are associated with ground-based employment at JSC.

In this era of increased sensitivity to patient confidentiality, it cannot be expected that the participating physicians will routinely share their suspicions without a mandate and mechanism to do so.

PARTICIPATION

Closely related to the ascertainment bias discussed in the previous section is the problem of participant drop out. Wear (2003) provided the committee with data on the percentage of participants appearing for scheduled physical exams each year from 1993 through 2001. Annual physical exams are required of active astronauts, so their participation rate is consistently near 100 percent. Table 3-1 shows the participation rates of active civil servants, comparison participants who are no longer working at JSC, and former members of the astronaut corps.

TABLE 3-3 Percentage of Active JSC Civil Servants, Ex-JSC Civil Servants, and Ex-Astronaut LSAH Participants Appearing for Scheduled Physical Exams, 1993-2001

	1993	1994	1995	1996	1997	1998	1999	2000	2001
Active JSC Civil Servants	93	79	87	79	79	84	84	82	69
Former JSC Civil Servants	72	53	61	67	59	65	59	63	50
Former Astronauts	71	69	61	64	61	88	63	74	68

Between 70 percent and 90 percent of active JSC civil servants receive their scheduled physical evaluations each year, but neither former JSC civil servants nor former astronauts regularly achieve a 70 percent return rate. Former astronauts have been returning at a slightly higher rate than former civil servants, at least in the 1998 to 2001 period.

Nine comparison participants have officially dropped out of the study, and the LSAH staff members have been unable to locate an additional five comparison participants. As noted briefly in the preceding chapter, it is possible that the rest of the study participants are all still "in the study" and a varying percentage simply miss an occasional physical exam. LSAH staff members were unable to provide data to evaluate that hypothesis, but it is their impression that this is not the case. Rather, the same individuals return over and over.

The most obvious explanation for the lower participation rates for the former civil servants and former astronauts is that many of them may have left the Houston area and find it inconvenient to return. No study of the reasons for nonparticipation has been performed, nor have any studies assessed remediable barriers to consistent participation. It may be that participants are more likely to return when healthy (and are able to travel easily) or when sick (and the physical exam may be of most use to them), but the travel requirement could introduce a significant bias. A further source of bias, which may explain why the former astronauts return somewhat more reliably than the former civil servants, is that the former astronauts are reimbursed for their travel to Houston, but the former civil servants are not. If travel is a significant factor, participation might be increased by arranging for or allowing participants to receive their examinations at a more convenient local site. The resulting increase in participation, if there is one, will have to be weighed against the additional variance introduced by inclusion of additional physician examiners.

An unrelated concern of the committee that may bear on participation stems from the copy of the informed consent form contained in the *LSAH Manual of Procedures* provided to the Institute of Medicine (IOM) staff. The committee recognizes, like the IOM committee which authored *Safe Passage* (IOM, 2001c), that the compulsory health surveillance of the active astronaut corps is occupational in nature and therefore not dependent on approval by the NASA institutional review board (IRB). However, inclusion of the data from active astronauts in the LSAH crosses the line into research activity, which does require the informed consent of the active astronauts. A vigorous explanatory effort should be part of the consent process.

Participation in the LSAH by former astronauts and all members of the comparison group is clearly covered by the Common Rule, and the *LSAH Manual of Procedures* has an appropriate section dealing with obtaining informed consent from these participants. However, the committee is concerned that the content and format of the consent form provided to them is not up to currently acceptable standards, specifically in regard to varying statements about the goals of the study, possible benefits, risks to confidentiality, compensation for participating, and procedures for encouraging participation. The committee believes that the risks to both confidentiality and full participation in the space program may not fall into the minimal risk category and encourages both the LSAH scientists and the NASA IRB to review the current informed consent process and make appropriate changes.

INTERNATIONAL PARTNERS

The charge to this committee included a request for recommendations on the inclusion of astronauts from NASA's international partners (Russia, Europe, Japan, Canada). The number of U.S. astronauts is still quite small by the standards of most epidemiological studies, especially when the health problems of greatest concern have a low incidence in the general population. For that reason, inclusion of all space travelers in the LSAH would seem advantageous. Inclusion of cosmonauts who spent months on the Mir space station would seem especially valuable. On the other hand, given the problems with the study already outlined above, as well as the reported differences among the partner countries in lifestyles, health care, mortality and morbidity, and expert opinion about what health measures and medical procedures are considered valuable or necessary in any circumstance, it is difficult to envision how space travelers from other countries could be enrolled in the LSAH without creating further difficulties in interpretation. Would JSC civil servants be appropriate comparisons for those astronauts and cosmonauts or would they need comparison participants from their own country and culture? If the latter, would the numbers ever be large enough to achieve the statistical power to detect any but the most obvious effects of space flight?

Negotiations among the international partners are already underway concerning the content, protocols, testing hardware and software, and data-sharing policies for the Clinical Status Evaluation (CSE). The CSE is to be a standardized battery of clinical, physiological, and psychological tests performed on each long-duration crew member from all International Space Station agencies. Pre-, in-, and post-flight evaluations will be used to guide clinical interventions, identify negative health or fitness trends in preclinical stages, plan optimal postflight rehabilitation, and assess the efficacy of countermeasures. NASA might find value in obtaining historical data from international partner astronauts and cosmonauts that is similar to that in the LSAH database. When and if CSE standards are finalized, the possibility of continuing regular CSE measurements in former International Space Station crews could be entertained.

FOLLOW-UP

The general question of NASA's responsibilities when the analysis of LSAH data suggests that astronauts may be at risk for a specific health problem is addressed in Chapter 5. The LSAH and the staff conducting it should not be generally held responsible for implementing preventive or therapeutic countermeasures, but it has and is playing major and very different roles in the two instances reported to date, cataracts and thyroid disorders. The committee therefore comments here on the actions taken in those two cases.

When available LSAH data showed an association between cataract incidence in astronauts (eye examinations for comparison participants is limited to visual acuity) and dose of radiation received during space flight radiation, the JSC scientific staff recognized the limitations of the study and contracted for a new study with more objective observation and classification of cataract severity and a comparison/control group of current and former military pilots age and gender matched to the astronauts. They also called for documenting lens opacities with digital photography for both astronauts and comparison participants in future LSAH physical exams and incorporation of cataract susceptibility into the debate on career radiation dose limits. The committee considers this approach a good general model, that is, whenever LSAH data show even a suggestion of a risk to astronauts (see the section earlier in this chapter on statistical power), the observation should be followed up with a more focused study using within-group comparisons based on such variables as time in space or radiation dose or by comparisons with a hypothesis-specific control group (brand new or a subset of the current comparison group).

In contrast, the committee believes that despite taking rapid and effective action to remediate the actual and potential problems of iodine-induced thyroid dysfunction by adding a filter to the space shuttle's drinking water taps, a proposed change in the medical standards restricting long duration space flight is unwise and premature. The proposal would exclude astronauts with thyroid autoantibodies or thyroid peroxidase antibodies from participating in long duration missions, including the International Space Station, because of concern about increased risk of thyroid dysfunction or even clinical thyroid disease with extended exposure to even low levels of iodine during long duration flight. There are insufficient data in the LSAH at present to fully assess these risks, but elevated thyroid autoantibodies occur about three times more frequently in women than in men and increasing in age (Eheman, 2003). The proposed exclusion would thus disproportionately limit the opportunity for female astronauts to participate on long duration missions.

The 16-fold reduction in drinking water iodine concentration produced by the anion exchange resin filters should eliminate or drastically reduce the risk of thyroid dysfunction in all future crew members. Additional filtering or adoption of an alternate water treatment regimen are obvious engineering solutions to that would protect all astronauts. Thyroxin (T4) replacement therapy would control these problems should they occur, but the International Board has previously argued against daily medications on missions, and the board, not JSC, controls the standards for international expeditions. Because prevention and treatment are so simple and effective, the IOM committee believes that it would be a mistake to adopt the proposed policy of excluding astronauts with thyroid autoantibodies or thyroid peroxidase antibodies from participating in long duration missions.

4

Recommendations for Changes in Study Design and Execution

The scientific and medical staff of the National Aeronautics and Space Administration (NASA) and the Wyle Laboratories contractors assisting them in managing the Longitudinal Study of Astronaut Health (LSAH) are well aware of the issues in study design and execution outlined in the previous chapter. In her presentation to the Institute of Medicine (IOM) committee in January 2003, Mary Wear in fact included their suggestions for correcting many of those deficiencies (Wear, 2003). This chapter will enumerate those suggestions and the committee's analysis of them, and then turn to additional recommendations from the committee itself.

LSAH STAFF RECOMMENDATIONS FOR CHANGE

LSAH scientists put forth suggested improvements to the study in three areas during their discussions with the IOM committee in March 2003: (1) improving the percentage of comparison participants returning to the Occupational Health Clinic at the Johnson Space Center (JSC) for physical exams, (2) improving the quality of the data collected from the comparison participants to provide a closer match to the data of the astronaut group, and (3) querying the database more frequently and systematically.

Improving the Quality of the Data Collected from the Comparison Participants

Budgetary restrictions have thus far prevented implementation, but LSAH staff members have proposed the following changes to the study:

- Offer the same physical examinations to the comparison participants as offered to the astronauts, on the same schedule.
- Include dual energy X-ray absorptiometry (DEXA) scans for the comparisons every third year.
- Offer the comprehensive profile of laboratory tests to all comparisons. Currently, only post-1992 comparison participants get the full profile, and then only every fourth year.
- Implement a more active program to obtain medical care records from private health care providers. The JSC Occupational Health Clinic provides no treatment for former employees. Participants are told the results of their physical exams and lab tests and referred to their private physicians for treatment of any suspected conditions. The current procedure is to simply ask participants to forward the records of those and any subsequent appointments.
- Offer to pay for an equivalent examination to be performed at a site more convenient to the participant.

Improving the Percentage of Comparison Participants Returning to the JSC Occupational Health Clinic for Physical Exams

LSAH staff members currently send a postcard to participants about a month before they are due for a physical exam (their birthday months), asking them to call the clinic and schedule an appointment. A second contact is made by letter or phone if the participant does not respond. If the participant indicates that he or she will be unable to return to JSC for an exam, the staff requests that the participant forward medical records for any visits to personal (i.e., non-JSC) health care professionals since the previous JSC clinic exam.

Staff proposes several inducements to improve "return rate," especially among retired civil service participants:

- Pay travel expenses for comparison participants who no longer work at JSC and live outside the Houston area. Former astronauts who live outside the Houston area are already reimbursed for travel expenses, as are active astronauts and JSC civil servant participants, if they incur any expenses. Although the LSAH staff discussed the desirability of providing some compensation for lost wages as well, they stopped short of including that in their recommended improvement.
- Offer to pay for an equivalent examination to be performed at a site convenient to comparison participants outside the Houston area. Occupational health clinics at other NASA centers, Federal Aviation Agency medical examiners, or private primary care providers could be given a standard protocol.

- Institute a publicity campaign to notify LSAH participants of the new benefit of receiving annual exams and laboratory tests comparable to those of the astronauts.
- Implement a more active program to identify and contact individuals who miss an annual physical.

Querying the Database More Frequently and Systematically

The LSAH staff proposes to hire another senior-level epidemiologist or statistician to perform statistical analyses, guide data analysts, and review results. At present the staff is completely consumed by the process of data collection. The current staff consists of a single doctoral level epidemiologist who also serves a the project manager, 2.5 masters level epidemiologists, an administrative assistant, 2 data entry clerks, and 3 software specialists.

IOM RECOMMENDATIONS FOR CHANGE

Implementing the following recommendations, which subsume many of those offered by the LSAH staff, will inevitably involve additional expenditures, but the committee believes they are essential for the validity of the data gathered through the LSAH and ultimately for the creation of a safer space travel program.

Recommendation 1

NASA should recognize that the LSAH can and should serve the two separate and potentially conflicting goals of occupational surveillance of the health of current and former astronauts and research into the long term health risks associated with manned space flight (and make these activities safer for future astronauts).

a. For the surveillance portion of the survey, participation of the astronauts is mandatory; for the research portion it is voluntary. Consequently, for the research portion, the astronauts need to sign an up-to-date informed consent document, and the research portion of the study should be reviewed on a regular basis by an IRB.

b. The database should be reviewed no less often than annually by LSAH staff, and analyses should be conducted for areas of potential risk, e.g., cancers, hearing loss, cataracts, bone strength. The committee is not convinced, given the low power of the study, that traditional "statistical significance" should be the sole trigger for concern, so in addition, it recommends that routine surveillance for unexpected and sentinel events be carried out by the oversight committee described below.

c. There should be a formal mechanism for flight surgeons to discuss both among themselves, and with those involved in the LSAH, any outlier or sentinel events, so that clinical suspicions are shared and checked for generality; such a system should complement the database surveillance system described above.

d. More information should be provided to participants on emerging findings and possible risks (possibly via their examining physician). The current newsletter system could be supplemented by a clinical synopsis with an expert commentary as key findings are published.

e. A formal process should be established to determine and implement corrective actions that follow from database surveillance or adverse event reporting. This process should enable the most learning to occur so that current and future astronauts are enabled to lead less risky lives, at least in their calling as explorers.

f. Review the Health and Lifestyle Questionnaire regularly with outside experts and update as recommended.

Recommendation 2

NASA should recognize that no comparison group can meet every goal of the LSAH. Although use of the existing comparison group can be improved (see below), other hypothesis-specific comparison groups will be needed for definitive assessment of specific risks identified in the astronaut population. The comparison group should be seen primarily as a means to detect possible anomalies. Only after anomalies are identified can the most appropriate control group be identified and a definitive assessment of risk made. Specific suggestions for the current comparison group are:

a) The ratio of three comparison participants for each astronaut selected should be maintained. JSC contractor (e.g., Wyle Laboratories) personnel should be added to the comparison participant pool if the civil servant population can no longer provide adequate matches for new astronaut classes;

b) NASA should continue to seek international partner astronauts' medical data, but we do not recommend pooling such data with the LSAH data;. Greater priority should be given to more thorough data gathering from the existing participant groups.

Recommendation 3

NASA should take steps to increase the quantity and improve the quality of the data collection and management of the data of the LSAH. The Committee was concerned by the marked variation in the content of the screening examinations that the existing LSAH groups (astronauts, retired astronauts, civil servants and retired civil servants) are currently

receiving, by the extent of missing data in some areas, and by the lack of justification for including some screening examinations and omitting others. These issues should be reviewed in accordance with the following principle: Exact or near-exact similarity of examination content in all four groups is more important than close similarity of examination frequency. Specific steps might include:

Data Collection

a) Pay travel expenses for comparison participants who no longer work at JSC and live outside the Houston area. Former astronauts who live outside the Houston area are already reimbursed for travel expenses, as are active astronauts and JSC civil servant participants, if they incur any expenses.

b) Offer to pay for an equivalent examination to be performed at a site convenient to comparison participants outside the Houston area. Occupational health clinics at other NASA centers, Federal Aviation Agency medical examiners, or private primary care providers could be given a standard protocol.

c) Institute a publicity campaign to notify LSAH participants of the new benefit of receiving physical exams and laboratory tests comparable to those of the astronauts.

d) Implement a more active program to identify and contact individuals who miss a scheduled physical and ascertain reasons for non-participation.

e) Implement a more active program to obtain medical care records from private health care providers. The JSC Occupational Health Clinic provides no treatment for former employees. Participants are simply told the results of their physical exams and lab tests and referred to their private physicians for treatment of any suspected conditions. Participants are asked to forward the records of those subsequent appointments with their private providers.

f) Inflight radiation dosimetry should be state of the art and carefully recorded in the LSAH database, along with exposures of both astronauts and comparison participants in diagnostic and therapeutic settings on earth. Analyses should be carried out by categories of "radiation dose" wherever possible.

The addition of the following would enhance the value of the study:

g) Mental health data should be added to the LSAH database.
h) Biological specimens should be stored for future tests and studies.

Data Management

The Committee recommends several changes in the oversight structure for the LSAH:

i) A standing oversight committee should be established with the participation of ex-astronauts, the public, scientists of various disciplines, and independent external reviewers. The expertise needed by the oversight committee includes such areas as biostatistics, environmental health, clinical medicine, radiation biology, neurology, endocrinology, and cardiology. Principal activities of such an oversight committee should be review of the methods used to acquire and analyze the data, surveillance of the data set for unexpected events, and evaluation of plans for reacting to these events. In addition, this oversight committee should set up procedures for site review of the performance of the study analogous to that performed by clinical research organizations.

j) At least one ex-astronaut and one or more non-NASA biomedical scientists should be added to the existing LSAH Executive Committee.

k) Additional professionals (e.g., epidemiologist) and staff should be hired as necessary to keep the database current and meet the new review and reporting requirements described above.

5

Recommendations for Changes in Health Care Policy

Whatever the goals, design, and execution of the Longitudinal Study of Astronaut Health (LSAH), the committee believes that the National Aeronautics and Space Administration (NASA) should have a policy addressing the practical consequences of discovering that a career as an astronaut, or the experience of space travel, leaves astronauts at increased risk for an adverse health effect. Of particular concern is the case in which the effect, cataracts for example, does not become obvious during or immediately after a space flight but instead develops sometime after the astronaut leaves active duty and is no longer provided medical care by NASA. What is NASA's ethical responsibility in this circumstance? It seems evident that the federal government should take full responsibility for health care needs in the case of a disease or disorder unique to space travel or the training required for space travel. Less clear is the case in which astronauts, as a group, are shown to be at increased risk for a common disease or disorder—skin cancer, perhaps. Any individual case may or may not be attributable to space flight. This chapter describes how other federal agencies have responded to similar discoveries of suspected occupational illness in former employees and ends with a recommendation on medical care for former astronauts.

DEPARTMENT OF ENERGY AND BERYLLIUM

Beryllium is a hard, gray metal that occurs as a chemical component of certain rocks (bertrandite and beryl), coal and oil, soil, and volcanic dust. Beryllium's light weight, high tensile strength, and ability to slow neutrons have made it useful for many purposes. Pure beryllium metal is used in the manufacture of aircraft disc brakes, nuclear weapons and reactors, missile parts, heat shields, X-ray machine parts, mirrors, and spacecraft. Beryllium oxide is used in ceramics for electronics and high technology applications, but from 1945 through the mid-1990s, more than 90 percent of all beryllium was processed for use by the Departments of Defense (DOD) and Energy (DOE) (and DOE's predecessor, the Atomic Energy Commission) to produce nuclear weapons.

Workers exposed to beryllium dust are at risk of developing serious, debilitating diseases. Acute (short-term) beryllium disease causes lung inflammation resembling pneumonia. Chronic beryllium disease (CBD) is a debilitating disease of the lung, apparently immunologically mediated. In severe cases, both the acute and chronic conditions may be fatal. The Department of Health and Human Services, Environmental Protection Agency, and International Agency for Research on Cancer consider beryllium to be carcinogenic.

Prevalence estimates of acute beryllium disease in Atomic Energy Commission (AEC) workers ranged as high as 7 percent in the late 1940s. An 8-hour average permissible exposure limit (PEL) of 2 micrograms/cubic meter was adopted in 1949 by the AEC, and it remained unchanged until 1999, when DOE lowered the PEL to 0.2 micrograms per 8-hour shift for its government workers and federal contractors. These standards were effective in eliminating most acute lung disease. CBD may have a latency of up to 30 years, however, and prior to the advent of current immunological tests, it frequently may have been misdiagnosed. Several studies of lung cancer in current and former beryllium workers returned the spotlight to beryllium hazards, and in 1993 Congress passed Public Law 102-484, which required the DOE to evaluate the long-range health conditions of current and former employees and contractors whose health might be at risk as a result of exposure to radioactive or other hazardous substances.

In a series of pilot studies throughout the 1990s, the DOE Former Workers Program established that it would be possible to locate and contact workers who might have been exposed to hazardous substances. DOE also began medical monitoring of current employees using the beryllium lymphocyte proliferation test (BeLPT). DOE estimates that about 20,000 current and former workers were or may have been exposed to beryllium. By the end of 1999 DOE had screened 13,770 of these workers and found 149 cases of CBD and an additional 299 workers with positive beryllium BeLPT tests but no clinical manifestations of disease.

In December 1999 DOE issued a rule establishing regulations to reduce beryllium exposure levels among its workforce, reduce the number of workers exposed to beryllium, and provide medical testing for exposed and potentially exposed workers. This rule on chronic beryllium disease prevention applies to federal, contractor, and subcontractor employees at 17 DOE facilities where there is actual or potential exposure to beryllium. In addition, the Secretary of Energy announced a legislative proposal reversing DOE's past practice of opposing and litigating most worker health compensation claims and providing compensation for employees who have contracted chronic beryllium disease or beryllium sensitivity.

The Energy Employees Occupational Illness Compensation Program Act of 2000 (Public Law 106-398) was enacted in October 2000, and the program went into effect on July 31, 2001. Administered by the U.S. Department of Labor, the program pays for medical monitoring of current and former DOE and DOE-contractor employees with one or more abnormal BeLPT tests. Individuals (or a surviving spouse, children, or parents) who develop CBD are eligible to receive $150,000 plus continuing coverage of costs for related medical care and treatment.

DEPARTMENT OF DEFENSE AND NUCLEAR WEAPONS TESTS

Between 1945 and 1963 the United States conducted more than 230 atmospheric tests of nuclear weapons. It is estimated that more than 200,000 DOD personnel, military and civilian, participated in these tests in some way. Some were merely witnesses; others set up scientific experiments and collected post-detonation data; still others participated in military exercises. Prescribed exposure limits varied as well but, as reported by Gladeck and Johnson (1996), they generally allowed maximum exposures of 3 to 5 rem (30 to 50 milliSieverts [mSv]) "per test or series." A series averaged 12 tests, but the Defense Threat Reduction Agency (DTRA), formerly the Defense Nuclear Agency (DNA), estimates that the average dose received by a participant was about 6 mSv (DTRA, 1999). This is approximately twice the average annual natural background dose received by a person living in the United States (NCRP, 1987). The DTRA (1999) study also estimates that less than one percent of all test participants received doses in excess of 50 mSv, the current dose limit for radiation industry workers.

Concerns among veterans about possible longterm effects on health persisted nonetheless, and when a veteran asserted in 1976 that his acute myelocytic leukemia was related to his participation in a 1957 nuclear test in Nevada, the Centers for Disease Control and Prevention conducted an epidemiological study of military personnel who attended that test. The study found more than the expected number of leukemia cases among them (Caldwell et al., 1980, 1983). In 1981 the Medical Follow-up Agency (MFUA) of the National Academy of Sciences began a study to evaluate the question of increased mortality among participants in other tests and at other locales. MFUA, in collaboration with DNA, studied 5 of the 19 U.S. atmospheric test series, incorporating tests over both land and sea, personnel from all branches of the armed forces (49,000 in all), and different kinds of nuclear devices. The mortality of veterans identified by DNA as having participated in at least one of the five selected test series was compared to mortality rates in the U.S. male population (Robinette et al., 1985). However, in 1989 DNA informed MFUA that it had misidentified many members of the participant cohort, potentially rendering the published findings invalid. At the request of DNA, MFUA redid the study, comparing the mortality of nearly 70,000 military personnel who participated in 1 of the 5 test series selected for the original study with the mortality of both the U.S. male population and the mortality of nearly 65,000 newly identified comparable veterans ("referents") who had not participated in any nuclear test (IOM, 2000). The study also benefited from an additional decade over which mortality could be observed. Results showed that test participants and referents had statistically similar risks of death from all causes, death from cancer, and death from leukemia. Although the difference was not statistically significant, leukemia mortality was 14 percent higher in the participant cohort (approximately 25 excess deaths). The authors chose to highlight this finding because of similar findings in two studies of military test participants in other nations, while discounting statistically significant increases in risk to participants of death from external causes (e.g., motor vehicle accidents) and nasal and prostate cancer as chance findings. The two supporting studies were on servicemen in the United Kingdom (Darby et al., 1993) and New Zealand (Pearce et al., 1996, 1997). The United Kingdom study found leukemia deaths were about 75 percent higher in 22,000 test participants than in 22,000 comparison personnel. The New Zealand study found leukemia mortality was elevated more than 400 percent among test participants, but it looked at only 528 test participants and 1,500 comparisons.

Various public laws as codified by Title 38, Code of Federal Regulations, Part 3, (38 CFR 3) authorize the Department of Veterans Affairs (VA) to provide medical care and to pay compensation benefits to confirmed test participants and dependency and indemnity compensation to certain survivors. Confirmed participants of U.S. atmospheric nuclear testing can receive special treatment for radiogenic diseases currently covered by VA regulation (38 CFR 3.309 and 3.311). Care for these conditions is provided without regard to the veteran's age, service-connected status, or ability to defray the cost of medical care. Additionally, no copayment by the veteran is required. Even if an eligible veteran has never filed a compensation claim or if the claim has been denied, the veteran can still receive free care for radiogenic diseases. Eligible veterans may also file a claim under either of both of the following compensation programs.

- *VA nonpresumptive program*: This program, codified in 38 CFR 3.311, provides for VA determination of service connection and benefits for about 25 specified disease categories, including any other cancer not specifically identified that are not presumed to be radiation-caused, but could be linked to radiation if the veteran's dose was high enough. These regulations define rules for adjudicating VA claims and establish a panel of experts to address scientific issues regarding the relationship between diseases and radiation. If a claimed disease is not one of the diseases listed in 38 CFR 3.311, the veteran must cite or submit competent scientific or medical evidence showing that there is a relationship between radiation exposure and the disease before consideration under this regulation may be made. To initiate a claim, veterans must submit to the VA competent medical evidence that the claimed medical condition exists. DTRA will respond to the VA's request by providing participation and associated radiation dose information based on dosimetry data for the individual or one or more other participants in the same military unit. A medical panel then determines, using organ-specific radiological tables derived from studies of Hiroshima and Nagasaki survivors, whether odds are equal to or greater than 0.5 that the individual veteran's condition is a result of test participation.

A recent examination of DTRA's dose reconstruction program by the National Research Council's Board on Radiation Effects Research found that very few (in the order of 50) claims have been granted despite a number of veteran-favorable assumptions in the dose reconstruction (NRC, 2003) The report attributes this result to the fact that ionizing radiation causes cancer only at significantly higher doses than those received by all but a few veterans of atomic weapons testing.

- *VA presumptive program:* Under this program, authorized by 38 CFR 3.309, the VA pays compensation for any of 21 types of cancers to eligible veterans without regard to radiation dose. To establish eligibility, veterans must submit competent medical evidence of the claimed medical condition to the VA. Upon receipt, the VA submits a request to DTRA to confirm the veteran's participation in U.S. atmospheric testing (or the occupation forces of Hiroshima and Nagasaki). Filing a VA claim under this law does not require dose information from any source; it *presumes* a connection between the veteran's participation in the test(s) and his current medical condition.

DEPARTMENT OF VETERANS AFFAIRS AND AGENT ORANGE

From 1962 to 1971 U.S. military forces sprayed more than 19 million gallons of herbicides over Vietnam to strip the thick jungle canopy that helped conceal opposition forces, destroy crops on which enemy forces might depend, and clear tall grass and bushes from around the perimeters of U.S. base camps and outlying fire-support bases. After a scientific report concluded that a contaminant in one of the primary chemicals used in the herbicide called Agent Orange could cause birth defects in laboratory animals, U.S. forces suspended use of the herbicide; they subsequently halted all herbicide spraying in Vietnam in 1971.

During the early and mid-1970s a growing number of veterans began to suggest a linkage between a variety of their diseases or conditions and their exposure to Agent Orange in Vietnam. A class action lawsuit filed in 1979 against five chemical manufacturers was eventually settled in 1985. It established a fund of $180 million to finance cash payments to totally disabled veterans and survivors of deceased veterans and set up an assistance foundation to help meet the medical, social, and legal assistance needs of members of the class.

The federal government also took a number of initiatives in response to concerns about the possible health consequences of wartime exposure to herbicides. Scientific studies were commissioned by both the executive and legislative branches, and Congress introduced several bills focused on health care and compensation for veterans exposed to Agent Orange. Public Law 97-72, enacted in November 1981, expanded eligibility for health care services at VA medical centers to include Vietnam veterans exposed to Agent Orange. The veteran is not required to demonstrate any link with Agent Orange, only proof of service in Vietnam. Care is provided unless the condition is shown to be due to some cause other than exposure.

Public Law 98-542, enacted in October 1984, addressed the issue of compensation for disabilities that may have resulted from exposure to Agent Orange in Vietnam. It provided for payment of disability and death benefits for Vietnam veterans with the skin diseases chloracne and porphyria cutanea tarda that became manifest within one year after service in Vietnam. The law also set up a mechanism for the VA to issue standards for determining disability claims based on exposure to Agent Orange. In brief, that mechanism calls for the Secretary of the Department of Veterans Affairs to presume a service connection (and provide for disability compensation where warranted) whenever there is sound scientific and medical evidence of a positive association between human exposure to an herbicidal agent and the occurrence of a disease in humans.

Public Law 102-4, the Agent Orange Act of 1991, extended disability compensation payments to Vietnam veterans for non-Hodgkin's lymphoma and several soft tissue sarcomas. This law also transferred the responsibility for determining the association between herbicide exposure and health outcomes from the VA to the National Academy of Sciences. A committee convened by the Institute of Medicine (IOM) of the National Academies conducted a major review of the scientific and medical evidence and in 1994 published a comprehensive report titled *Veterans and Agent Orange: Health Effects of Herbicides Used in Vietnam* (IOM, 1994). Periodic updates have followed (IOM, 1996, 1999, 2000, 2002, 2003a).

The original IOM committee approached its task by assigning each of the diseases and disorders under study to one of four categories on the basis of the epidemiological evidence reviewed:

1. Sufficient evidence of an association;
2. Limited or suggestive evidence of an association;
3. Inadequate or insufficient evidence to determine whether an association exists;
4. Limited/suggestive evidence that *no* association exists.

The committee's *Update 2002* (IOM, 2003a) lists four health outcomes in the first category: soft tissue sarcoma, Hodgkin's disease, non-Hodgkin's lymphoma, and chloracne. In the second category (limited or suggestive evidence of an association) are respiratory cancers of the lungs/bronchus, larynx, and trachea; prostate cancer; multiple myeloma acute and subacute transient peripheral neuropathy; porphyria cutanea tarda; type 2 diabetes; and spina bifida in children of veterans. Twenty-four other health outcomes have been classified as having inadequate or insufficient evidence to determine whether an association with herbicides exists, and two further outcomes (brain tumors and gastrointestinal tumors) fell in the category of limited/suggestive evidence of no association.

The second element of that committee's charge was to determine the increased risk of disease among people exposed to herbicides during service in Vietnam. The committee and its successors have reviewed numerous studies of the health of Vietnam veterans but to date have found it impossible to adequately quantify the risk to veterans, largely because of uncertainty about the nature and magnitude of exposures in Vietnam. A recent report on a modeling effort based on temporal and spatial proximity to known herbicide spraying operations encourages some additional efforts in this regard (IOM, 2003b).

The result of these two products of the IOM's Agent Orange committees has been the decision by the VA to provide compensation for Vietnam veterans for all the conditions in the first two categories listed above without requiring evidence of exposure other than service in Vietnam.

CONCLUSIONS

Some elements of these examples may be useful in formulating a NASA policy on its responsibilities to astronauts, despite some critical differences among them and between each of them and the situation facing NASA. Former beryllium workers with CBD or lung cancer receive treatment and compensation without evidence of a causal link in each case. The value of the beryllium case is limited by the specificity of the causal agent in the case of CBD, and the near-zero incidence of the disease outside workplaces involving beryllium. Space-flight-specific diseases may yet emerge as the LSAH accrues more data, but at the moment it appears more likely that NASA authorities will be faced with the more difficult problem of determining whether individual cases of relatively common diseases (cataracts, for example) in former astronauts are occupation- or spaceflight-related.

Vietnam veterans with a wide variety of maladies have argued that their conditions were due to exposure to Agent Orange in the course of their duties. As was the case for DOE and former beryllium workers, the VA lacked good records of exposure. Unlike the DOE case, in which there was not much question that beryllium was a necessary condition for CBD, the VA faced claims from veterans with a myriad of diseases found not only in Vietnam veterans but also in individuals without known exposure to Agent Orange and similar herbicides. Considerable resources were expended establishing a short list of 11 disorders for which there is at least suggestive evidence of an association with herbicide exposure. Having established these links, the government provides compensation for Vietnam veterans without requiring further evidence of a causal link in each case.

In the case of nuclear test participants, VA had more information available about exposure, in the form of radiation badges for a subset of participants, and as a result of extensive follow up of Hiroshima and Nagasaki survivors, a body of data about which cancers were radiation-related and a dose-response curve that described that relationship. For some of these cancers and other health problems, compensation is provided only if the test participant's documented dose meets a liberal but nevertheless specified threshold. In most cases, however, documentation of one of 15 different cancers and participation in a nuclear test is sufficient for compensation.

In each of the three examples, Congress has authored legislation to ensure that the federal government errs on the side of finding too many of its former employees eligible for care and compensation. Given the high profile of the NASA astronaut corps, it is hard to imagine a radical change in that approach should the LSAH reveal that former astronauts have an elevated risk for certain medical problems. Given the small size of the astronaut corps and their unique service to the country, the committee believes that NASA should take a page from military retirement policy and assume responsibility for the lifelong health care of former astronauts. Not only would it be the right thing to do, but it would preempt both adverse public opinion and Congressional intervention. An associated benefit to NASA might be increased participation in the LSAH.

The committee's final recommendation, therefore, reiterates a suggestion of their predecessors on the committee that authored *Safe Passage* (IOM, 2001c).

Recommendation 4

NASA should assume responsibility for the lifelong health care of its active and former astronauts.

References

Band PR, Spinelli JJ, Ng VT, Moody J, Gallagher RP. 1990. Mortality and cancer incidence in a cohort of commercial airline pilots. *Aviation Space and Environmental Medicine* 61(4):299–302.

Caldwell GG, Kelley D, Heath CW Jr. 1980. Leukemia among participants in military maneuvers at a nuclear bomb test: A preliminary report. *Journal of the American Medical Association* 244(14):1575–1578.

Caldwell GG, Kelley D, Zack M, Falk H, Heath CW Jr. 1983. Mortality and cancer frequency among military nuclear test (Smoky) participants, 1957 through 1979. *Journal of the American Medical Association* 250(5):620–624.

Cucinotta FA. 2003 (January 15). *Study of Cataracts in Astronauts.* Paper presented at meeting of the Institute of Medicine Committee on Aerospace Medicine and Medicine in Extreme Environments, Cape Canaveral, FL.

Cucinotta FA, Manuel FK, Jones J, Iszard G, Murrey J, Djojonegro B, Wear M. 2001. Space radiation and cataracts in astronauts. *Radiation Research* 156(5 Pt 1):460–466.

Darby SC, Kendall GM, Fell TP, Doll R, Goodill AA, Conquest AH, Jackson DA, Haylock RGE. 1993. *Mortality and Cancer Incidence 1952-1990 in UK Participants in the UK Atmospheric Nuclear Weapons Tests and Experimental Programmes.* Oxon, UK: National Radiological Protection Board.

DTRA (Defense Threat Reduction Agency). 1999. *Radiation Exposure in U.S. Atmospheric Nuclear Weapons Testing, (Nuclear Test Personnel Review Program).* [Online]. Available: http://www.dtra.mil/news/fact/nw_ntprpre.html.

Eheman CR, Garbe P, Tuttle RM. 2003. Autoimmune thyroid disease associated with environmental thyroidal irritation. *Thyroid.* 2003;13(5):453-64.

Fischer CL. 2003 (January 15). *Cancer Incidence in the U.S. Astronaut Population.* Paper presented at meeting of the Institute of Medicine Committee on Aerospace Medicine and Medicine in Extreme Environments, Cape Canaveral, FL.

Fox AJ, Collier PF. 1976. Low mortality rates in industrial cohort studies due to selection for work and survival in the industry. *British Journal of Preventive and Social Medicine* 30: 225-230.

Gladeck F, Johnson, A. 1996. *For the Record: A History of the Nuclear Test Personnel Review Program 1978-1993.* Defense Nuclear Agency: DNA-6041F.

Hamm PB, Billica RD, Johnson GS, Wear ML, Pool SL. 1998. Risk of cancer mortality among the Longitudinal Study of Astronaut Health (LSAH) participants. *Aviation Space and Environmental Medicine* 69(2):142–144.

Hamm PB, Nicogossian AE, Pool SL, Wear ML, Billica RD. 2000. Design and current status of the Longitudinal Study of Astronaut Health. *Aviation Space and Environmental Medicine* 71(6):564–570.

IOM (Institute of Medicine). 1994. *Veterans and Agent Orange: Health Effects of Herbicides Used in Vietnam.* Washington, DC: National Academies Press.

IOM. 1996. *Veterans and Agent Orange: Update 1996.* Washington, DC: National Academies Press.

IOM. 1999. *Veterans and Agent Orange: Update 1998.* Washington, DC: National Academies Press.

IOM. 2000. *The Five Series Study: Mortality of Military Participants in U.S. Nuclear Weapons Tests.* Washington, DC: National Academy Press.

IOM. 2001a. *Exploring the Biological Contributions to Human Health: Does Sex Matter?* Washington, DC: National Academies Press.

IOM. 2001b. *Veterans and Agent Orange: Update 2000.* Washington, DC: National Academies Press.

IOM. 2001c. *Safe Passage: Astronaut Care for Exploration Missions.* Washington, DC: National Academies Press.

IOM. 2002. *Veterans and Agent Orange: Herbicide/Dioxin Exposure and Acute Myelogenous Leukemia in the Children of Vietnam Veterans.* Washington, DC: National Academies Press.

IOM. 2003a. *Characterizing Exposure of Veterans and Agent Orange and Other Herbicides Used in Vietnam: Interim Findings and Recommendations.* Washington, DC: National Academies Press.

IOM. 2003b. *Veterans and Agent Orange.* Washington, DC: National Academies Press.

McMonigal KA, Braverman LE, Dunn JT, Stanbury JB, Wear ML, Hamm PB, Sauer RL, Billica RD, Pool SL 2000. Thyroid function changes related to use of iodinated water in the U.S. Space Program. *Aviation Space and Environmental Medicine* 71:1120-5.

McMonigal KA. 2003 (March 31). *Thyroid Disease and Exposure to Iodine During Space Flight.* Paper presented at meeting of the Institute of Medicine Committee on Aerospace Medicine and Medicine in Extreme Environments, Washington, DC.

Nicogossian AE, Huntoon CL, Pool SL. 1993. *Space Physiology and Medicine.* 3rd ed. Philadelphia: Lea and Febiger.

NAS (National Academy of Sciences). 1967. *Radiological Factors in Manned Space Flight.* Washington, DC: National Academies Press.

NRC (National Research Council). 1980. *The Effects on Populations of Exposure to Low Levels of Ionizing Radiation (BEIR III) Final Report.* Washington, DC: National Academies Press.

NRC. 1990. *Health Effects of Exposure to Low Levels of Ionizing Radiation (BEIR V).* Washington, DC: National Academies Press.

NRC 2003. *A Review of the Dose Reconstruction Program of the Defense Threat Reduction Agency.* Washington, DC: National Academies Press.

Pearce N. 1996. *Mortality and Cancer Incidence in New Zealand Participants in United Kingdom Nuclear Weapons Tests in the Pacific: Supplementary Report.* Wellington, NZ: Wellington School of Medicine.

Pearce N, Winkelmann R, Kennedy J, Lewis S, Purdie G, Slater T, Prior I, Fraser J. 1997. Further follow-up of New Zealand participants in United Kingdom atmospheric nuclear weapons tests in the Pacific. *Cancer Causes and Control* 8:139–145.

Peterson LE, Pepper LJ, Hamm PB, Gilbert SL. 1993. Longitudinal Study on Astronaut Health: Mortality in the years 1959-1991. *Radiation Research* 133:257–264.

Robinette CD, Jablon S, Preston TL. 1985. *Mortality of the Nuclear Weapons Test Participants.* Washington, DC: National Academies Press.

Thompson DE, Mabuchi K, Ron E, Soda M, Tokunaga M, Ochobuko S, Sugimoto S, Ikeda T, Terasaki M, Izumi S. 1994. Cancer incidence in atomic bomb survivors. Part II: Solid tumors. *Radiation Research* 137(suppl.):S17–S67.

Wear M. 2003 (January 15). *The Longitudinal Study of Astronaut Health.* Paper prepared for meeting of the IOM Committee on Aerospace Medicine and Medicine in Extreme Environments, Cape Canaveral, FL.

Wolff J, Chaikoff IL. 1949. The temporary nature of the inhibitory action of excess iodide on organic iodine synthesis in the normal thyroid. *Endocrinology* 45:504–513.

A

Biographical Sketches of Committee and Staff

DAVID E. LONGNECKER, M.D., (chair) is senior vice president and corporate chief medical officer and the Robert Dunning Dripps Professor of Anesthesia at the University of Pennsylvania. Dr. Longnecker received his M.D. in 1964 from Indiana University School of Medicine, where he completed residency training in anesthesiology in 1967. Following a National Institutes of Health (NIH) Special Research Fellowship in Physiology, he continued clinical and laboratory research at the NIH Clinical Center, where he served as a clinical associate from 1968 to 1970. He has received numerous NIH research grants and a Research Career Development Award for research involving the effects of anesthetics on microcirculation, oxygen delivery to tissue, oxygen therapeutics, endothelium-dependent circulatory control, and health services research. Dr. Longnecker is the author and coauthor of more than 175 scientific abstracts and original scientific articles, 29 chapters, and five textbooks. Dr. Longnecker was previously the Harold Carron Professor of Anesthesiology at the University of Virginia and, from 1970 to 1973, assistant professor of anesthesiology and physiology at the University of Missouri. Dr. Longnecker is a member of the Institute of Medicine (IOM) and previously chaired the IOM Committee on Fluid Resuscitation for Combat Casualties.

ALFRED F. CONNORS, JR., M.D., is currently the chair of the Department of Medicine at Case Western Reserve University's MetroHealth Medical Center. Prior to accepting that position he was director of the Division of Health Services Research and Outcomes Evaluation at the University of Virginia, where he also acted as coclinical director of the Medical Management Program and associate director of the Center for Minority Health. A graduate of St. Louis University, Dr. Connors received his medical degree from the Medical College of Ohio. He is board-certified in internal medicine, critical care medicine, and pulmonary diseases. Dr. Connors previously taught at Case Western Reserve University, where he received the Kaiser Teaching Excellence Award as Best Clinical Teacher in 1991. He serves on the editorial boards of three prestigious journals, and is the author of more than 100 professional publications, including a book on differential diagnosis and 15 book chapters. His current research is supported by grants from the Department of Defense, Department of Veterans Affairs, several institutes at the National Institutes of Health, and Agency for Health Research and Quality.

ROY L. DEHART, M.D., M.P.H., is director of the Vanderbilt Center for Occupational and Environmental Medicine. Dr. DeHart completed medical school and his internship at the University of Tennessee College of Medicine, Memphis, and his residency in aerospace medicine at Johns Hopkins University and the United States Air Force (USAF) School of Aerospace Medicine. He is board certified in aerospace medicine and occupational medicine. In his assignment as chief of aerospace medicine for Air Force Systems Command, Dr. DeHart maintained an active role with the National Aviation and Space Administration. He served as a medical monitor of manned space flights as well as on a number of research panels addressing future manned space operations. As a senior aerospace medical officer he commanded the Air Force's Aerospace Medical Research Laboratory at Wright Patterson Air Force Base, Ohio; his final assignment was as commander of the USAF School of Aerospace Medicine. After 23 years in the Air Force, he shifted his activities to the field of occupational medicine and became interested in the phenomena known as multiple chemical sensitivity. He is a senior aviation medical examiner for the Federal Aviation Administration and sees both private and commercial pilots for their medical certification. He currently serves on the Presidential Advisory Board on Radiation and Worker Health. He is the recipient of numerous honors and awards, the most recent being the Knudson Award presented by the American College of Occupational and Environmental Medicine. Dr. DeHart is the editor of three editions of the textbook *Fundamentals of Aerospace Medicine*, which has been described by reviewers as the premier text in the discipline.

R. J. MICHAEL FRY, M.D., is retired from Oak Ridge National Laboratory, where he was head of the cancer section of the biology division. He continues to serve as consultant to the life sciences division of the laboratory and as professor at the University of Illinois. He holds M.B., B.Ch., and M.D. degrees from the University of Dublin, Ireland. Previous appointments include as senior scientist in the division of biology and medicine at Argonne National Laboratory and professor in the Department of Radiology, University of Chicago. He was editor of *Radiation Research* and continues as a consulting editor. His research focused on the effects, in particular, carcinogenesis, of ionizing and ultraviolet radiation. He is a member of the International Committee on Radiation Protections Committee 1; member of the National Council on Radiation Protection's Committees 1, 4, 7, 75, and 88; and chairman of the National Research Council's Board on

Radiation Effects Research. His awards include the Failla and Hartman medals and the Lauriston Taylor lectureship.

DANIEL MASYS, M.D., is director of biomedical informatics and professor of medicine at the University of California San Diego School of Medicine. An honors graduate of Princeton University and the Ohio State University College of Medicine, he completed postgraduate training in internal medicine, hematology, and medical oncology at the University of California, San Diego, and the Naval Regional Medical Center, San Diego. Previously, he served as chief of the International Cancer Research Data Bank of the National Cancer Institute, National Institutes of Health, and from 1986 through 1994 was director of the Lister Hill National Center for Biomedical Communications. In this capacity, Dr. Masys served as the chief program architect and first director of the National Center for Biotechnology Information established within the National Library of Medicine in 1987 to support molecular databases and computational tools. Dr. Masys is a diplomate of the American Board of Internal Medicine in Medicine, Hematology, and Medical Oncology. He is a fellow of the American College of Physicians, fellow of the American College of Medical Informatics, and member of the Institute of Medicine. He has served as a consultant to the NASA Life Science Informatics program and is an active, instrument-rated pilot.

VAN C. MOW, Ph.D., is Stanley Dicker Professor of Biomedical Engineering and Orthopaedic Bioengineering and chairman of the Department of Biomedical Engineering at the Engineering School of Columbia University and director of the New York Orthopaedic Hospital Research Laboratory at the College of Physicians and Surgeons of Columbia-Presbyterian Medical Center. Dr. Mow earned an undergraduate degree in aeronautical engineering and a Ph.D. in applied mechanics from Rensselaer Polytechnic Institute. He has edited six monographs on orthopedic biomechanics, diarthrodial joint function, cell mechanics, and cellular engineering. He has published more than 650 full-length papers, book chapters, and extended abstracts. He is currently co-editor of the journal *Osteoarthritis and Cartilage* and founder of the *Journal of Orthopaedic Research*. Among his many awards and honors are membership in both the National Academy of Engineering and Institute of Medicine. Dr. Mow is currently a member of both the Committee on Membership and the Russ Prize Committee of the National Academy of Engineering. Currently, he is also a member of the International Steering Committee for the Fourth World Congress of Biomechanics and chairman of the Advisory Committee for the Division of Medical Engineering Research, National Health Research Institute, Taiwan.

TOM S. NEUMAN, M.D., is professor of medicine and surgery and associate director of the department of emergency medical services at the University of California San Diego Medical Center. A graduate of Cornell University, he received his M.D. from the New York University School of Medicine in 1971, followed by his internship and residency in internal medicine at Bellevue Hospital. Dr. Neuman is board certified in internal medicine, pulmonary disease, occupational medicine, and emergency medicine. He is a fellow of the American College of Physicians and the American College of Preventive Medicine. Dr. Neuman has been a leader in the field of the physiology and medicine of diving throughout his career and was the editor-in-chief of *Undersea and Hyperbaric Medicine* until July 2002. He is the co-editor of the most widely used textbook of diving medicine and physiology. He previously served on the IOM Committee on Space Medicine.

THOMAS F. OLTMANNS, Ph.D., is the Edgar James Swift Professor of Arts and Sciences in the department of psychology at Washington University in St. Louis, Missouri. He previously served as professor of psychology and psychiatric medicine and director of clinical training in psychology at the University of Virginia. He has also served as professor of psychology at Indiana University. Dr. Oltmanns received his undergraduate degree from the University of Wisconsin, Madison, and his Ph.D. at the State University of New York at Stony Brook. He has authored 5 books and more than 50 journal articles. Dr. Oltmanns is past president of the Society for a Science of Clinical Psychology and is a consulting editor for the *Journal of Abnormal Psychology* and a member of the editorial board for *Psychological Bulletin* and *Journal of Personality Disorders*. His research has been supported by numerous grants, and he is currently co-principal investigator on a large grant looking at peer assessment of personality traits and pathology. He has served on two different grant review committees for the National Institute of Mental Health and is a member of NASA's Astronaut Selection Psychiatric Standards Working Group.

RUSSELL B. RAYMAN, M.D., M.P.H., is executive director of the Aerospace Medical Association in Alexandria, Virginia, a position he assumed in 1992 after a long and distinguished career in the U.S. Air Force and a brief interlude as manager of medical operations for Lockheed Engineering and Sciences Company. Among his many positions in the Air Force were commander of two different hospitals, chief of the medical readiness division in the Office of the Surgeon General, and consultant in aerospace medicine to the Assistant Secretary of Defense for Health Affairs. He has held academic appointments at the University of Texas, San Antonio, Uniformed Services University of the Health Sciences, Wright State University, and University of Texas Medical Branch. Dr. Rayman has published more than 50 papers and chapters and 2 books and was a member of the National Academy of Sciences Committee on Air Quality in Passenger Cabins of Commercial Aircraft. He holds a medical degree from the University of Michigan and a masters of public health from Johns Hopkins University, and is board-certified in both family practice and aerospace medicine. He is also certified in aviation medicine by the Royal College of Physicians (London). He completed the National Aeronautics and Space Agency flight surgeon certification course in 1989 and serves on the agency's aerospace medicine and occupational health advisory committee.

WALTER M. ROBINSON, M.D., M.P.H., is associate professor of pediatrics and medical ethics at Harvard University. After receiving his M.D. from Emory University in 1988, Dr. Robinson spent two years at Boston City Hospital and one year at Johns Hopkins as a pediatric resident. Following one year of work in a neighborhood health center, he returned to a fellowship in pediatric pulmonary medicine at Harvard/Children's Hospital and also served a fellowship in medical ethics. He received an M.P.H. from Harvard in 1994. Dr. Robinson is board-certified in pediatrics and pediatric pulmonary medicine. He has continued to practice while teaching ethics at the medical school and postgraduate levels and serving on committees appropriate to his expertise. He directs the Harvard Ethics Fellowship Program and serves on the editorial board of *Ethics and Behavior*. He previously served as a member of the IOM Committee on Space Medicine.

ELAINE RON, Ph.D., is a senior scientist and former chief of the Radiation Epidemiology Branch in the Division of Cancer Epidemiology and Genetics at the National Cancer Institute. She holds an M.P.H. in health service administration from Yale University and a Ph.D. in epidemiology from the Tel Aviv University. Her research focuses on the carcinogenic effects of radiation exposure and the epidemiology of thyroid cancer. Dr. Ron is a member of Committee 1 of the International Commission on Radiological Protection (ICRP). She also has served as a consultant to the United Nations Scientific Committee on the Effects of Atomic Radiation (UNSCEAR) and as an advisor to the NCRP Committee 1-8 on Induction of Thyroid Cancer by Ionizing Radiation. Dr. Ron was a member of the National Academy of Sciences Committee to study the mortality of military personnel present at atmospheric tests of nuclear weapons. She is a fellow of the American Epidemiology Society and a member of the American Thyroid Association and the Radiation Research Society.

CAROL SCOTT-CONNER, M.D., Ph.D., is head of the department of surgery at the University of Iowa, Iowa City. Dr. Scott-Conner received her M.D. from the New York University School of Medicine in 1976 and stayed for her surgical residency, which she completed in 1981. After leaving NYU, she joined the faculty at Marshall University and then moved to the University of Mississippi. During her tenure she earned a Ph.D. in anatomy from the University of Kentucky and an M.B.A. from Else School of Management at Millsaps College. Since 1995 she has been a professor and the head of surgery at the University of Iowa. Dr. Scott-Conner is board-certified in surgery and in surgical critical care and is a fellow of the American College of Surgeons. Dr. Scott-Conner has been active on 22 editorial boards, and has authored 5 books and more than 200 original papers, abstracts, reviews, and book chapters. She holds memberships in many elected surgical societies and has frequently served in leadership positions. She received her undergraduate degree in electrical engineering at the

Massachusetts Institute of Technology. She previously served as a member of the IOM Committee on Space Medicine.

RHEA SEDDON, M.D., is the assistant chief medical officer for the Vanderbilt Medical Group in Nashville, Tennessee, and a former three-flight veteran astronaut for NASA. As an astronaut, she logged over 722 hours in space. She was a mission specialist on STS-51D and STS-40 and was the payload commander on STS-58. Dr. Seddon also served in several other capacities at NASA, namely as technical assistant to the director of flight crew operations and special advisor for Shuttle/Mir scientific payloads and as a member of NASA's Aerospace Medical Advisory Committee and the International Bioethics Task Force. After earning a B.A. in physiology at the University of California, Berkeley, and an M.D. from the University of Tennessee, Dr. Seddon went on to complete an internship and residency in general surgery in Memphis. Dr. Seddon's areas of interest are in emergency medicine and nutrition.

DEBORAH R. ZUCKER, M.D., Ph.D., is assistant professor of medicine at Tufts University School of Medicine, clinical investigator in the Department of Medicine's Division of Clinical Care Research, and a practicing internist in the General Medical Associates clinic at New England Medical Center. Dr. Zucker's research focuses on developing methods to enable widespread practice-based clinical research through the combination of individual patient (N-of-1) studies and is currently evaluating treatments for fibromyalgia. She has also focused on a range of clinical and health policy topics including acute sinusitis, gender, and the presentations of acute myocardial infarction, physicians' conflicts of interest in managed care, and programs to curb antibiotic resistance. She was assistant director of the New England Medical Center-Agency for Healthcare Research and Quality (AHRQ) Evidence Based Practice Center and coinvestigator for an AHRQ-sponsored program to train evidence-based practitioners. After receiving her Ph.D. in microbiology and molecular genetics from Harvard University, Dr. Zucker received her M.D. from the University of Miami School of Medicine. She has completed postgraduate training in internal medicine, medical ethics, and health services research.

IOM Board on Health Sciences Policy Staff

FREDERICK J. MANNING, Ph.D., is a senior program officer in the IOM's Board on Health Sciences Policy and study director. In nine years at IOM he has served as study director for projects addressing a variety of topics, including medical isotopes, potential hepatitis drugs, blood safety and availability, rheumatic disease, resource sharing in biomedical research, occupational safety and health, and chemical and biological terrorism. Before joining IOM, Dr. Manning spent 25 years in the U.S. Army Medical Research and Development

Command, serving in positions that included director of neuropsychiatry at the Walter Reed Army Institute of Research and chief research psychologist for the Army Medical Department. Dr. Manning earned a Ph.D. in psychology from Harvard University in 1970, following undergraduate education at the College of the Holy Cross.

ANDREW POPE, Ph.D., is director of the Board on Health Sciences Policy at the Institute of Medicine. With expertise in physiology and biochemistry, his primary interests focus on environmental and occupational influences on human health. Dr. Pope's previous research activities focused on the neuroendocrine and reproductive effects of various environmental substances on food-producing animals. During his tenure at the National Academy of Sciences and since 1989 at the Institute of Medicine, Dr. Pope has directed numerous studies on topics that include injury control, disability prevention, biologic markers, neurotoxicology, indoor allergens, and the enhancement of environmental and occupational health content in medical and nursing school curricula. Most recently, Dr. Pope directed studies on priority-setting processes at the National Institutes of Health, fluid resuscitation practices in combat casualties, and organ procurement and transplantation.

MELVIN H. WORTH, JR., M.D., is a scholar-in-residence at the Institute of Medicine. Dr. Worth completed his surgery residency at New York University-Bellevue in 1961 and remained on that faculty for 18 years. He founded the Bellevue Trauma Service in 1966 and continued as director until 1979, when he left to become director of surgery at Staten Island University Hospital. He served for 15 years with the New York State Office of Professional Medical Conduct and 8 years as a member of the New York State Hospital Review and Planning Council (for which he was chair in 1993). He is a fellow of the American College of Surgeons, American College of Gastroenterology, and International Society for Surgery and holds memberships in the American Association for the Surgery of Trauma, Society for Critical Care Medicine, Association for Academic Surgery, New York Surgical Society (for which he was president in 1979), and other academic and professional organizations. Dr. Worth retains his appointment at New York University and is clinical professor of surgery at the State University of New York Downstate (Brooklyn) and the Uniformed Services University of the Health Sciences. Dr. Worth most recently served as an IOM study staff member to the Committee on Fluid Resuscitation for Combat Casualties and is the senior advisor to the Committee on Creating a Vision for Space Medicine During Travel Beyond Earth Orbit.

BENJAMIN N. HAMLIN, B.A., Research Assistant at the Institute of Medicine, received his bachelors degree in biology from the College of Wooster in 1993 and a degree in health sciences from the University of Akron in 1996. He then worked as a surgeon's assistant in the fields of vascular, thoracic, and general surgery for several years before joining the National Academies staff in 2000. As a Research Assistant for the Division on Earth and Life Studies at the National Academies, Ben worked with the Board on Radiation Effects Research on projects studying the health effects of ionizing and non-ionizing radiations on the human body. He has worked on a number of Institute of Medicine studies, including Testosterone and Aging: Clinical Research Directions; Review of NASA's Longitudinal Study of Astronaut Health; Health Literacy: A Prescription to End Confusion; Improving Medical Education: Enhancing the Behavioral and Social Science Content in Medical School Curricula, and NIH Extramural Center Programs: Criteria for Initiation and Evaluation. Ben is currently pursuing graduate work in the sociomedical sciences. He is also involved with the U.S.-Bangladesh Advisory Council, an organization that promotes governmental cooperation between the United States and Bangladesh on matters of trade and healthcare.

NATASHA S. DICKSON has been a senior project assistant with the National Academies' Institute of Medicine since March 2001. She is a graduate of the John S. Donaldson Technical Institute in Trinidad and Tobago. She gained administrative experience at the University of the West Indies, St. Augustine, and also worked as an advertising sales representative and reporter for the Trinidad Express Newspapers.

B

Variables in the LSAH Database

Physical Examination

Vital signs: Sitting, standing and recumbent blood pressure and pulse, height, weight, percent body fat (based on sum of four skinfolds), temperature
Clinical evaluation: Review of systems, significant interval history, summary of defects and diagnoses, recommendations, qualifying information
Dental exam results
Vision (full exam): Corrected and uncorrected distant and near vision, each eye and binocular; color vision, depth perception, heterophorias, intraocular pressure; only near and distant acuity tested for comparison participants
Audiometry: 500, 1,000, 2,000, 3,000, 4,000, 6,000, 8,000 Hz for each ear
ECG: Consultant interpretation
Pulmonary function test: Standard spirometry test
Dual energy X-ray absorptiometry (DEXA) scan
Exercise tolerance test (85% max): Bruce Protocol - heart rate and blood pressure recorded under the following conditions: every minute for 5 minutes in supine position, every minute for 3 minutes standing, immediately after running in place and every minute for 3 minutes, every minute for 3 minutes at each of the following grade/mph: 10/1.7, 12/2.5, 14/3.4, 16/4.2, 18/5.0, 20/5.5, 22/6.0, and during recovery every minute until return to baseline, total exercise time, reason test terminated early
Chest X-ray interpretation
Proctosigmoidoscopy interpretation
Colonoscopy interpretation
Mammogram consultant interpretation
Pelvic exam/Pap smear; pathology report

Comprehensive Laboratory Analysis

Hematology: WBC, RBC, hemoglobin, hematocrit, MCV, MCH, MCHC, platelet count, RDW, reticulocyte count, differential (neutrophils, lymphocytes, monocytes, basophils, eosinophils)
Chemistry panel: Glucose, BUN, uric acid, creatinine, total bilirubin, aspartate transaminase, alanine transaminase, alkaline phosphatase, lactate dehydrogenase, glutamyltransferase, sodium, potassium, chloride, phosphorus, calcium, magnesium, carbon dioxide, total protein
Serum iron: Iron, ferritin, total iron binding capacity, transferrin, transferrin saturation
Lipid profile: Cholesterol, triglyceride, VLDL, HDL, LDL, Chol/HLD ratio
Urinalysis: pH, specific gravity, color, appearance, protein, glucose, ketone, blood, bilirubin, urobilinogen, nitrite, leukocyte esterase, WBC, RBC, epithelial cells, mucus
Ionized calcium profile: Ionized calcium, ionized calcium at 7.40
SPE panel: Total protein, albumin, alpha 1, alpha 2, beta, gamma, A/G ratio
Immunoglobulin panel: IGG, IGA, IGM, IGE
Serology: Hepatitis A total, hepatitis B surface antigen, hepatitis B surface antibody, hepatitis C antibody, RPR, CRP, anti-HIV
Thyroid function tests

Personal Medical History

Self-report of personal medical history, checklist review of medical problems, hospitalizations

Medical Records Obtained from the Johnson Space Center Medical Clinics, Private Physicians and Hospitals

Acute and chronic medical events, diagnoses, medical procedures, treatment, medications, recommendations

Lifestyle Questionnaire

Marital status and history, family medical history, reproductive history, smoking history, alcohol use, exercise and weight patterns, pilot experience, hormone use (women only)

Death Records

Death certificate, hospitalization records, and autopsy reports, if available, are obtained

Post-Flight Medical Debrief (astronauts only)

Twenty-five page questionnaire/interview: self-report of medical events that occurred during flight, inflight exercise, medications use, quality of sleep, habitability issues, recovery symptoms post-flight

C

Health Lifestyle Questionnaire

LONGITUDINAL STUDY OF ASTRONAUT HEALTH

Manual of Procedures – Revision Date: 6/1/98

Chapter 4: Data Management

Section 4.1.3.8.1

LSAH08a Form – LSAH Health Lifestyle Questionnaire

The Longitudinal Study of Astronaut Health (LSAH) Health Lifestyle Questionnaire, shown on the following pages, was developed in 1993 to collect needed data that were not consistently collected (or not collected at all) by other instruments. Starting in 1993, this form has been sent to participants to complete upon their selection into the study, and is completed by the participants.

Longitudinal Study of Astronaut Health Health
Lifestyle Questionnaire

This questionnaire is intended to obtain demographic information, family medical history, and personal health risk factors which are not otherwise obtained in a systematic manner for all LSAH participants. If you do not have, or if you choose not to provide, some of the information requested in this questionnaire, please skip that question and continue to the next question. All LSAH data are protected under the Privacy Act and will be reported as group data only.

1. Last Name: ______________________________

2. First Name: ______________________________

3. Middle Initial: ______

4. Maiden Name: ____________________________

5. Date Form Completed: ______________________

Demographic Information

6. Education (check highest level achieved):

__ High school diploma or equivalent
__ Technical or vocational certification
__ Some college
__ College degree(s)

List all undergraduate and graduate degrees (B.A.., M.S., Ph.D., etc.) on the line below:

____

7. What is your current marital status (circle only one choice)?
   1. Never Married 2. Married 3. Divorced
   4. Separated 5. Widowed

8. If ever married, please complete for each marriage:

	If Marriage Ended			
Year (date) of Marriage	Year (date) Marriage Ended	Marriage ended with (circle one):		
		1. Divorce	2. Death	3. Other
		1. Divorce	2. Death	3. Other
		1. Divorce	2. Death	3. Other
		1. Divorce	2. Death	3. Other
		1. Divorce	2. Death	3. Other
		1. Divorce	2. Death	3. Other

9. What is your race or ethnic group (circle only one choice)?
   1. African American
   2. Asian
   3. Hispanic
   4. Native American
   5. White
   6. Other (please specify) ___________________________

Family Medical History

10. Are your biological parents still living?
    Father: ___ Yes ___ No ___ Unknown
    Mother: ___ Yes ___ No ___ Unknown

	If Living		If Deceased		
	Age	Serious Illnesses/ Conditions (current or past)	Age at Death	Specific Cause(s) of Death	Serious Illnesses/ Conditions While Alive
Mother:					
Father:					

11. Do you have any brothers or sisters (alive or deceased)

___ Yes ___ No

If "yes," how many (total number)? _____

If no, go to question #13

12. Please complete for each brother and sister, alive or deceased (if you need more space to write, use the blank sheet of paper attached to the end of this questionnaire).

If Living			**If Deceased**			
Sex	Age	Serious Illnesses/ Conditions (current or past)	Sex	Age at Death	Specific Cause(s) of Death	Serious Illnesses/ Conditions While Alive

Reproductive History

13. Have you or your partner(s) ever been pregnant (circle one choice)? Please answer using the information below. If MALE, report the pregnancy(ies) of your partner(s), while she(they) were your partner(s). IF FEMALE, report your own pregnancies.

1. Yes 2. No (if no, go to question # 16)

How many times have you or your partner(s) been pregnant (include miscarriages, stillbirths, and abortions; if pregnant now, include the pregnancy in this answer)?
___ pregnancies

14. Please complete the chart below for your reproductive history, but do not include current pregnancy.

Pregnancy #	Pregnancy Ended in (check one):				Date of Birth or Date Pregnancy Ended			Duration of Pregnancy			Sex of Child			Congenital Birth Defects, Deformities or Disorders Present at Birth			Cur-rently Living		If Living, Does Child Have Any Chronic Illnesses or Conditions			If Deceased, Date of Death			If Deceased, Describe Cause(s) of Death
	Live Birth	Still Birth	Miscarriage	Abortion	Month	Day	Year	Weeks	Months	Unknown	Male	Female	Unknown	Yes	No	If yes, describe	Yes	No	Yes	No	If yes, describe	Month	Day	Year	
1																									
2																									
3																									
4																									
5																									
6																									
7																									
8																									
9																									
10																									

Stress-related Information

15. How satisfied are you with life (circle only one choice)?
    1. Not satisfied 2. Somewhat satisfied 3. Mostly satisfied
    4. Totally satisfied

16. Compared to last year, how satisfied are you with life (circle only one choice)?
    1. Less 2. The same 3. More

17. Have you experienced PLEASANT life events in the past 12 months (circle only one choice)?
    1. Many 2. Some 3. Few 4. None

18. Have you experienced UNPLEASANT life events in the past 12 months (circle only one choice)?
    1. Many 2. Some 3. Few 4. None

19. In general, would you say your current state of health is (circle only choice):
    1. Excellent 2. Very good 3. Good 4. Fair 5. Poor

20. Please enter the number of SICK DAYS, i.e., days that you stayed home due to personal illness, in the past 12 months, (include weekends and holidays). _____ Days

21. Please enter the number of days that you were HOSPITALIZED in the past 12 months. Do not include hospital stays of less than 24 hours.
    _____ Days

Smoking History

22. Which of the following best describes your use of cigarettes (circle only one choice)?
    1. Never smoked – less than 100 cigarettes (go to question # 26)
    2. Ever smoked (current or past smoker)

23. If you ever smoked, how old were you when you started to smoke on a regular basis?
    _____ Years of age

24. What is your current smoking status? *Circle only one choice and answer the questions listed with that choice.*
    1. Current Smoker:
       How many cigarettes do you smoke per day?
       _____ Cigarettes per day (go to question # 26)

    2. Past Smoker:
       a. How many years has it been since you smoked cigarettes on a regular basis?
          _____ Years

       b. On average, how many cigarettes per day did you smoke before you quit?
          _____ Cigarettes per day

25. Do you currently use smokeless tobacco (such as chewing tobacco or snuff)?
    1. Yes 2. No

Alcohol Use

26. In the past two weeks, on how many days did you drink alcoholic beverages such as beer, wine, or liquor (circle only one choice)?
    1. Do not drink (go to question # 29)
    2. Sometimes drink but none in past 2 weeks (go to question #29)
    3. 1 to 3 days
    4. 4 to 6 days
    5. 7 to 10 days
    6. 11 to 14 days

27. On the days that you drank, how many drinks did you have per day on the average, (circle only one choice)? One drink = 12 oz. beer; 1 oz. liquor; or 4-6 oz. wine.
    1. One drink 2. Two drinks 3. Three to four drinks
    4. Five or more drinks

Aspirin Use

28. Do you routinely take aspirin (do <u>not</u> include nonaspirin pain relievers, i.e., Tylenol, Ibuprofen, etc.)?
    1. Yes 2. No (go to question # 32)

29. How many aspirin do you usually take in a single dose?
    _____ Tablet(s) of adult dosage (325 mg) aspirin
    _____ Tablet(s) of children's dosage (80 mg) aspirin

30. How frequently do you take aspirin (circle only one choice)?
    1. Multiple doses per day
    2. One dose per day
    3. 2 to 6 doses per week
    4. 1 dose per week
    5. Fewer than 1 dose per week

Exercise and Weight Patterns

31. Which of the following best describes your exercise routine (circle only one choice)?
    1. Exercise every day
    2. Exercise regularly; 5 to 6 times per week
    3. Exercise regularly; 3 to 4 times per week
    4. Exercise regularly; 1 or 2 times per week
    5. Exercise occasionally
    6. Do not exercise at all (go to question # 35)

32. If you exercise, what is the normal length of your exercise periods (circle only one choice)?
    1. Less than 20 minutes
    2. More than 20 minutes, but less than 1 hour
    3. 1 to 2 hours
    4. More than 2 hours

33. Which of the following types of exercise most closely describes your usual exercise (check all that apply)?
    ___ Running or jogging
    ___ Swimming
    ___ Bicycling
    ___ Walking
    ___ Sports (i.e. soccer, basketball, softball, football, etc.)
    ___ Racquet sports
    ___ Weight lifting or resistance training
    ___ Rowing
    ___ Skating (roller or ice)
    ___ Aerobics/dance/calisthenics

34. During the past year, has your weight fluctuated more than 5 pounds (circle only one choice)?
    1. Yes 2. No (go to question # 40)

35. If "yes," which of the following patterns best describes the weight changes you experienced during the past year (circle only one choice)?
    1. Only lost weight
    2. Only gained weight
    3. Gained, then lost in one cycle
    4. Gained, then lost in more than one cycle
    5. Lost, then gained in one cycle
    6. Lost, then gained in more than one cycle

36. What was the maximum gain you experienced in the past year? _____ Pounds

37. What was the maximum loss you experienced in the past year? _____ Pounds

38. Did you intentionally change your weight?
    1. Yes 2. No

Pilot and Nonpilot Flight Experience

39. Are you a licensed pilot?
    1. Yes 2. No

40. If "yes," please record the number of flight hours you logged in the past year for each type of aircraft listed below.

Type of Aircraft	Hours Logged
1. High performance aircraft	_____
2. Jet aircraft – more than 30 seats	_____
3. Jet aircraft – fewer than 31 seats	_____
4. Helicopter	_____
5. Propeller – more than 30 seats	_____
6. Propeller – fewer than 31 seats, but more than 4 seats	_____
7. Small aircraft with fewer than 4 seats	_____
8. Glider	_____

41. As a passenger (commercial or otherwise), or in any other nonpilot capacity, do you fly more than 100 hours per year (do <u>not</u> include hours logged as a pilot)?
    1. Yes 2. No

42. If "yes," approximately how many hours did you fly during the past year (do <u>not</u> include hours logged as a pilot)?
    1. 100-200 hours 2. 200-300 hours 3. 300 or more hours

For Women Only: Hormone Usage

43. Have you ever taken birth control pills or used other hormonal birth control methods (i.e., Depo-provera, Norplant)?
    1. Yes 2. No (go to question # 46)

44. If "yes," what is the total number of years you have used hormonal birth control?
    _____ Years

45. Have you stopped having your menstrual periods?
    1. Yes 2. No (go to question # 50)

46. If "yes," how old were you when you stopped having menstrual periods?
    _____ Years of age

47. Did you have surgical intervention which ended your menstrual periods (hysterectomy or removal of ovaries)?
    1. Yes 2. No

48. If you had surgical intervention, did you have both ovaries removed?
    1. Yes 2. No

49. Have you ever taken estrogen replacement therapy?
    1. Yes 2. No

50. If "yes," how many years in total have you taken estrogen replacement therapy?
    _____ Years

--- END ---

Thank you for completing this questionnaire.